Praise For This Book

"A careful, clear, and absolutely devastating portrait of our folly!"

~ Bill McKibben, Founder of 350.org, author, *Eaarth*

"This brilliantly-written book commands our attention to the challenges we face as sea level rises at rates unknown to modern humankind. Englander manages to capture the grand sweep of sea level rise over geologic history while also presenting projected future changes in ways that are very convincing, and daunting. It is a pleasure to recommend this very readable and persuasive book."

~ Dr. Robert W. Corell, lead author, U.S. National Climate Assessment; former assistant director for geosciences, National Science Foundation

"'High Tide on Main Street' is a courageous book that looks squarely in the face of a problem far greater than most of us dare to imagine. John Englander's writing is clear and accessible, employing vivid metaphors and avoiding technical jargon, making it easy to grasp a fascinating scientific story of urgent importance for our time."

~ Dr. Ben Strauss of Climate Central

"Englander provides us a fair warning. Disastrous collapse of the Greenland and West Antarctic ice sheets can be avoided only if the public pressures our ... leaders to rapidly phase out fossil fuels."

~ Dr. James Hansen, renowned climatologist; former Director of NASA's Goddard Institute of Space Studies

"Read this book... the clearest statement to date about the frightful reality of accelerating sea level rise and the catastrophic impacts... a crisis beyond anything civilization has encountered. Englander shows what we must begin doing now to slow this change, and examines the adaptation options we have for our communities, countries and earth."

~ Dr. Harold R. Wanless, Chair of Department of Geological Sciences, University of Miami

"If you own coastal property or know someone who does, you better read this now! The time coastal property values start collapsing is nigh."

~ Joseph Romm, Senior Fellow, Center for American Progress; Editor, ClimateProgress.org

"An excellent read. This book provides a synthesis of the state-of-the-art scientific understandings of sea level— past, present and future, and paints a clear picture of the implications for us and for future generations."

~ Dr. Ralph Rayner, Editor in Chief, *Journal of Operational Oceanography*; Professor, London School of Economics

"We owe it to ourselves to read this book and learn all that we can to be actively engaged in a sustainable future. With knowing comes caring, with caring comes hope and with hope the possibility of a better future."

~ Jean-Michel Cousteau President of Ocean Futures Society

"John Englander provides a clear picture of climate change with an emphasis on sea level rise and its implications for our future. If you live on or near a coast or if you enjoy the beach, this is a must read for a deeper appreciation of the damages being done by our fossil fuel addiction."

~ Larry Schweiger, President National Wildlife Federation

"Sea level rise has important implications for the wide range of industries that depend on ports and coastal infrastructure. 'High Tide on Main Street' brings sea level rise into focus now, for the coming decades, and beyond that will be critical for ocean business community planning and adaptation."

~ Paul Holthus, Executive Director, World Ocean Council

"In the current debate on climate change, or rather climate destabilization, relatively little is usually said about the inexorable rise in sea levels. Here short term remedies are impossible. In 'High Tide on Main Street' John Englander spells out in accessible, often personal style, the science, the risks, the impacts, and the long-term implications."

~ Sir Crispin Tickell, former President, Royal Geographical Society (UK)

"Well done merging of science, policy and human behavior. This book accurately portrays the science, along with the mitigation and adaptation options."

~ Dr. Larry Atkinson, Professor of Oceanography, Old Dominion University

"An easy-to-read, but disturbing analysis of the impacts of sea level rise now and way into the future. Compulsory reading for all who live on or near the world's coasts."

~ Dr. Leonard Berry, Director, Florida Atlantic University's Center for Environmental Studies

"John Englander has made a complex subject accessible and interesting. 'High Tide on Main Street' illuminates inspirational examples of how various communities and organizations are facing this challenge and converting it into real opportunities to enhance our community and economic resilience."

~ Margaret A. Davidson, Acting Director, NOAA's Office of Ocean and Coastal Resource Management

"John Englander's 'High Tide on Main Street' presents sea level rise from a global interdisciplinary perspective that makes clear the economic, political, legal, and other challenges we face in beginning the necessary adaptations now."

~ Richard Hildreth, Kliks Professor of Law and Director, University of Oregon Ocean and Coastal Law Center

"...First time anyone has clearly explained that greenhouse gases are not just accelerating climate change, but have actually changed the natural ice age cycles of the last three million years, and why that should matter "

~ Daniel Kreeger, Executive Director, Association of Climate Change Officers

High Tide on Main Street

High Tide on Main Street

Rising Sea Level and the Coming Coastal Crisis

by John Englander

The Science Bookshelf

Published by:
The Science Bookshelf
P.O. Box 652
Boca Raton, FL 33429-0652

Second Edition, 2013
Revision 2.1, 2014

For errata & revision log, see
www.hightideonmainstreet.com/revisions

Published in the United States of America
ISBN 978-0615637952
Library of Congress Catalog # 2012917567

Cover design by Kata Jancso, katajancso@yahoo.com

All truths are easy to understand once they are discovered;
the point is to discover them.

~ Galileo

Contents

To my parents, Bernard and Ruth, for their love and guidance.

To my daughter, Rachel, whose love guides my concern
for future generations.

Foreword: Jean-Michel Cousteau

It may be no surprise that my entire life has focused on the sea. My father, Jacques Cousteau, and my mother Simone made the sea part of my life from my earliest recollections as a child. Even with all I have learned about the ocean, this book has added a valuable new perspective.

High Tide on Main Street is fascinating and should be essential reading for anyone wanting to understand the impact of rising seas, which is happening much sooner than most realize. The very real threat of rising sea level, compounded by the increasing concentration of population, cities, and valuable infrastructure on the coasts worldwide makes this a truly important work. Hundreds of millions of people are vulnerable and hardly anyone understands the facts or the consequences.

John has done an admirable job of assembling the current science and explaining it in a way that anyone can grasp. This is a very critical message for home owners, political leaders, and anyone running businesses or with personal real estate interests near the coast. The impacts are rather astonishing. The insights into the economic consequences and our options for addressing this issue are clever and novel.

The timing is also poignant. Recently, we commemorated the 100th anniversary of the birth of my father, Jacques-Yves Cousteau. It is but a footnote of history that months before his death in 1997, he asked John Englander to take over as CEO of The Cousteau Society, of which I was a founder. My father had spent several days with John, read his articles about the state of the ocean, and witnessed him communicating to his peers in the scuba diving industry about the facts and trends affecting the sea. Within weeks my dad and John had agreed that he would become CEO, with my dad remaining as chairman. Many of the old crew of the *Calypso* and I shared great hope that this turn of events would bring his skills to that organization. Unfortunately, John's time as CEO was shortened due to my father's untimely death.

With that background, however, it is particularly exciting now to see this book as a clear voice about the ocean. It proves what my father saw in John more than a decade ago—that he, too, is an objective observer, that he is knowledgeable about the sea, and that he communicates well.

John has been a close friend and colleague for decades, both personally and with my organization, Ocean Futures Society, which I created in honor of my father's philosophy. I truly applaud his first book. Our ocean planet, which we all depend upon, desperately needs more voices that can explain the science in terms that are relevant beyond the scientific community, and inspire us all to be a part of a more sustainable future. How can we protect what we do not fully understand? With this book, more and more people will understand the science of rising sea levels and will want to protect our planet and ultimately protect ourselves.

J.M.C.
Santa Barbara, California
2012

Preface

Hurricane Sandy hit the New York City region on October 29th, 2012, just one week after the first edition of this book was published. The timing was eerie as the book described exactly that type of storm hitting exactly that location. Tempting though it is now, one year later, I have decided not to rewrite the book to cover Sandy in retrospect. The devastation is well known and gives extreme validation to the scenario I describe starting on the second page of the first chapter, in many places throughout and explicitly in chapter twelve. This second edition allows me to clarify numerous points that have been brought to my attention by the thousands of readers. Rather than just focus on Sandy, this is an opportunity to look at additional city vulnerabilities, incorporate the latest-2013 IPCC report, and provide a little deeper coverage about our options moving forward. By leaving the relevant sections on page 121 about New York City intact, I hope to preserve how it foreshadowed real events far sooner than I ever could have imagined.

Standing on the rocky coast of Greenland in 2007 at dusk, sipping some extraordinary scotch, I suddenly had clarity for this book. I was CEO of an exemplary nonprofit organization, The International SeaKeepers Society. Mostly composed of yacht owners, the group equipped vessels to record precise ocean and atmospheric measurements. As the ships sailed, data was transmitted by satellite for use by scientists. It was a clever concept, yielding almost free data for the public good.

Members of SeaKeepers were with me in Greenland, ground zero for the melting ice sheet and glaciers. Dr. Robert Corell, a leading Arctic scientist, was our guide. The next day, he would take us to Jakobshavn Glacier, the largest in Greenland, and the one that oceanographers were quite certain spawned the iceberg that sank the *Titanic.*

And that's when it hit me. Among all the confusing aspects of climate change, the reality of sea level rise alone might get the public's attention enough that they recognize what is at stake. Amid all the issues of climate change, the profound and permanent threat of sea level rise is barely appreciated. Outside of the circles of geologists and oceanographers, hardly anyone realizes the vast range of sea level rise. Yet, most people have some connection to the coast. The eight people accompanying me to Greenland lived in Denver, and rising sea level even had their interest.

Despite all the confusion, the uncertainties in the models, and the complexity, I could explain one clear message:

> *As sea level rises, the shoreline will move hundreds, even thousands of feet inland, destroying vast amounts of property, including most coastal communities. The eventual economic impacts will be absolutely enormous, and may begin to be felt within a decade. The change will last for centuries, eventually reaching many tens of feet higher than now. It will be the first time this has occurred in more than 100,000 years.*

Currently, the vast majority of people are simply unaware of the truth about rising sea level. Ignorance, intentional disinformation, and a lack of leadership have certainly played a role. I also believe that short-term thinking and wishful thinking are part of the explanation.

We all enjoy opportunities to dream, to envision the future we wish for. Such dreaming is a powerful and valuable vehicle for setting goals and effecting change. But to have tangible effect, our dreams must be based in reality. Wishful thinking by itself will not solve difficult problems. We need facts and feasible plans to deal with the challenges we face.

Over the next decade or two, our largely coastal-oriented civilization is about to get a real-world lesson in geology and oceanography. Rising sea level will have catastrophic, far-reaching effects for many, many generations. I hope this book helps you fully appreciate what lies ahead and better prepare for it.

J.E.
Boca Raton, FL
Amended October 2013

Introduction: Governor Christine Todd Whitman

In late October 2012, Hurricane Sandy devastated my home state of New Jersey. Dozens of lives were lost, and thousands of homes and businesses destroyed. For many, the recovery will take years. Fortunately, we are a strong and resilient state.

Sometime in that swirling aftermath, I heard about this book that prophetically described just such an event hitting the region from Atlantic City to New York. Even more eerily, it had been published exactly one week before Hurricane Sandy made landfall.

In "High Tide On Main Street", John Englander describes a new era of sea level rise combining with severe storms and extreme tides to fundamentally reshape our shorelines. This is a rare book that has excellent scientific credentials, yet speaks in plain language and uses memorable visualizations to paint a clear picture.

As a former Governor, and former Administrator of the US Environmental Protection Agency, my immediate tendency is to look at this vision in terms of public policy. Yet I also know that this is something that should be understood by every community, company, and citizen. This book provides a very helpful explanation of a world that has just started to change and will continue for decades far into the future.

It is a privilege to introduce this second edition, published on the first anniversary of that landmark event. Please read this book and share it. If enough people see the long-term forces now at work, we can have an enlightened discussion that will allow us to better prepare for extreme storms and better adapt to a changing shoreline. We owe it to ourselves and to future generations.

C.T.W

Tewksbury Township, New Jersey
2013

The Big Picture

Chapter One
Aboard the *Titanic*

When all the ice sheets and glaciers in the world melt, sea level will be approximately 212-feet (65-meters) higher than it is today.[1] That paralyzing fact is independent of any confusion about climate change. It has happened before and will happen again.

Satellite images and field observations provide irrefutable evidence that we are now experiencing the beginnings of what will be massive, long-term melting. From the frozen Arctic and Greenland to the perimeter of Antarctica, ice is disappearing at a quickening pace. While it is not deemed possible for the full meltdown to happen this century, or even next, the realities of huge sea level change need to be understood because they affect everything on the coasts worldwide.

Rising sea level will be the single most profound geologic change in recorded human history. It will transform our physical world beyond anything we can imagine, dwarfing continents and eliminating some nations. Coastlines will move inland by hundreds and, in some places, thousands of feet this century. The impacts will be far greater during the next century. Trillions of dollars of the most valuable real estate and infrastructure will vanish.

Some of the changes will happen incrementally, and barely be noticeable within a year. With increasing frequency, however, extreme high tides, storms, and erosion will change shorelines suddenly and dramatically. Either way, the decline in coastal real estate values will be immense, with far-reaching implications.

While the issue of sea level rise is starting to draw concern, the focus tends to be on the accuracy of short-term forecasts while missing the larger truth that sea level will almost certainly rise for at least 1,000 years.

The topic of sea level rise also gets lost in the larger, more complicated discussion of climate change. There is an obvious connection in that a warmer planet will create higher sea levels. But somehow the discussion has become a debate over whether it is caused by human activity or is simply the result of natural cycles, with most people believing it's one or the other.

In an age where media spin prevails, anyone can look for their preferred idea and support their argument with what seems like solid evidence. This phenomenon might be amusing if the stakes weren't so high. Over the course of this century, hardly a moment in geologic time, there will be the proverbial train wreck—or perhaps that should be shipwreck. It will be the unavoidable collision between a rising sea and our civilization, much of which is located on the coasts. The effects of rising sea level are already happening and will continue to worsen.

The band-aid solutions of restoring beaches, building seawalls, subsidizing flood insurance, and rebuilding flooded cities may seem like worthy approaches, but only if your time frame is very short. We deny the inevitable long-term consequences at truly great cost to us, and even more so to future generations.

This issue has profound moral and ethical implications since it affects many innocent others, including our heirs. Our grandchildren will not likely look back kindly on our era if we do not quickly begin to expand the scope of our response beyond what is politically and financially expedient.

This may seem like a bleak forecast, but the success of our species is intricately linked to our ability to adapt. If we have the courage to look at our future, we can make it a livable one.

Using the story of the *Titanic* as a metaphor, we are just emerging from the fog bank and now know what lies ahead. We too are heading for a collision of sorts, in our case between the relentlessly rising ocean and our densely developed coastal society.

As passengers on our "ship", we might be categorized by a range of responses similar to the passengers who were aboard the *SS Titanic*. Many will just "party" until catastrophe, effectively wanting to ignore the bad news; a large number will be immobilized by the news; others will focus anger on the cause; some will work to minimize impact; and others will plan for the aftermath of the inevitable.

As we consider which attitudes will characterize our own personal response, we need to understand what is going to happen in greater detail. Throughout this book, we will explore:

1. How much sea level could rise by the middle of this century and beyond.

2. The factors that could increase or slow future sea level rise.

3. How far inland shorelines will move as sea level rises, and what the impacts will be.

4. How soon coastal property values could begin to drop in anticipation of the property eventually going underwater, and what the implications will be for individuals, businesses, the banking system and government at all levels.

5. How we can begin "intelligent adaptation."

While this drama will unfold over decades, the early signs of change can already be seen around the world. The years 2011 and 2012 brought headlines from Seattle, Miami, Norfolk, and San Diego in the United States, and internationally from Italy, China, and Bangladesh, where streets flooded during above-average tides, known as "king tides." Few have thought to ask why this new phenomenon is happening and what it implies for the future.

In the year since the first edition was published, multiple headlines seem to have been pulled from its pages, notably starting with an event like Hurricane Sandy hitting New York City, to ominous changes in one particular place in West Antarctica, to a crisis with the US National Flood Insurance program.

As sea level rises, today's king tide is tomorrow's routine high tide. Daily high tides, monthly lunar tides, and intermittent storm surges are all moving steadily higher. It is important to grasp the magnitude and inevitability of this while there is time to plan and adapt.

Chapter Two
Earth's 612-foot Elevator

It's no wonder that rising sea level is confusing and hard to accept. We often hear in vague terms that sea level has been higher and lower in the past, without any clear explanation of the scope and scale of natural cycles. Meanwhile, for most of recorded human history, sea level has barely changed at all and there are still few obvious signs of rising seas. Topping it off is the fact that you might have heard projections for sea level rise this century ranging from seven inches to more than 20 feet. What is true and what is hype?

To get perspective let's look at a few major eras of sea level when it was far higher and far lower than the present. Figure 2-1 shows that 500 million years ago, sea level was much higher than it is now. Over the last few million years, it has risen and dropped hundreds of feet, with the current sea level height being toward the top of typical cycles.

One of the aspects of climate change that is hard to see is annual variations compared to longer term trends. Even the longer trends are not necessarily linear. The trend can move along slowly by the decades and then hit a "step" either a sharp up turn, or even a plateau phase, and then repeat. These are counter-intuitive, not what we would expect. This non-linearity applies to sea level and to temperature as we will explore.

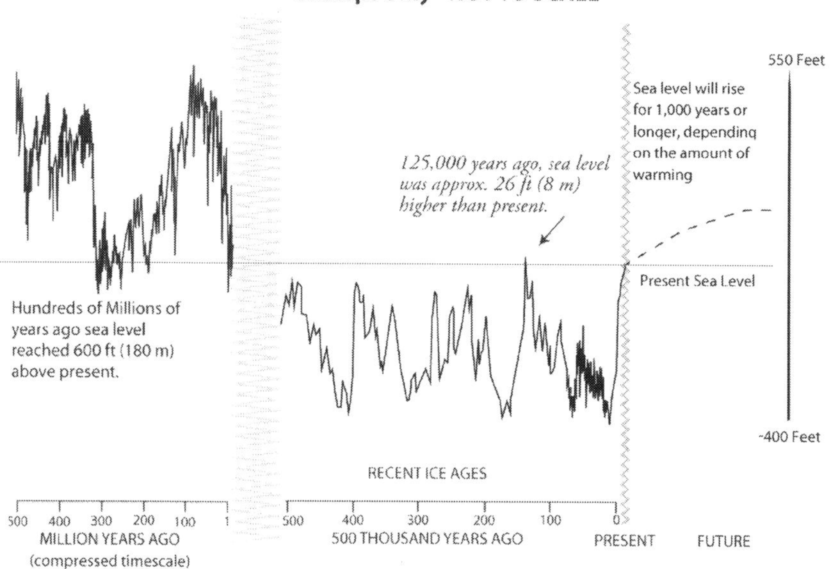

Figure 2-1. Sea level has changed dramatically during Earth's long history, as shown in this concept diagram. Hundreds of millions of years ago, when the planet was warmer and the ice sheets were smaller or nonexistent, sea level was much higher. For the last few million years sea level has moved up and down by 300-400 feet, following the ice ages. Now sea level has started a new phase of rising, which will last for centuries, or millennia, due to the ice sheets melting for the first time in millions of years. (Graphic by John Englander. May be reproduced with author citation.)

Fluctuations in sea level can occur for many reasons, but fundamentally sea level change is a function of the average long-term temperature of the atmosphere and the ocean. The warmer the temperature, the smaller the ice sheets and glaciers, and the higher the sea level. Our understanding of sea level is one side of a coin; climate change is the other.

Setting aside the issue of humans' effect on climate for the moment, it should be understood that geoscientists recognize a few different types of climate change:

1. **Abrupt climate changes** can be precipitated by external events such as meteorite impacts, extraordinarily large volcanic eruptions, and "tipping points," when the accumulation of massive amounts of powerful greenhouse gases have caused significant changes in temperature.

2. **Regular, periodic climate changes** are measured in cycles ranging from one-year to 100,000-year ice age cycles. Annual seasonal cycles are driven by planetary phenomena such as the tilt of the earth on its axis and its elliptical orbit around the sun, which determines the amount of heat received each season.

3. **Dramatic changes in earth systems dynamics**, such as the moving plates of the earth's crust. In the shorter term, plate tectonics can create earthquakes and volcanic activity as the plates grind and shift on the order of inches or feet at a time. Over the long term—millions of years—the movement of continents can create profound climate changes. For example, about 50 million years ago, one big plate jammed into present-day Eurasia. We now know that plate as the Indian subcontinent. Over the millions of years that the plate moved across the shallow, carbonate-rich Indian Ocean, it stirred up sediments, greatly increasing atmospheric carbon dioxide levels to around 1,000 parts per million (ppm), causing substantial warming.[2] When that Indian plate smashed into Eurasia, it pushed up a large ridge of land that we now call the Himalayas. The rise of that towering mountain range had a direct impact on air currents, temperatures, and global weather patterns.[3]

WHERE WE ARE IN THE CYCLE

After the peak of the last ice age, around 20,000 years ago, the earth shifted to a warming phase, putting the ice sheets back into melt mode. During the next 14,000 years, sea level rose almost 400 feet. Then, for roughly the past 6,000-8,000 years, the normal cycles of huge changes in ocean height stopped and sea level remained steady. We were in a plateau phase.

Over the last few centuries, the long-term warming phase was in the very slow and chaotic process of changing direction to enter the next cooling phase with its corresponding falling sea level. However, burning enormous amounts of fossil fuel in the last 100 years has changed the level of carbon dioxide from the pre-industrial high of roughly 280 ppm to nearly 400 ppm in 2012, and climbing.

The resultant warming is now overpowering the natural cooling force, and the rise in temperature is quickly accelerating sea level rise, mostly observed in the rapidly-melting ice sheets and glaciers in Greenland and Antarctica. The level of warming and ice melting is well outside the pattern of the last few million years.

An Elevator With 47 Floors

As stated in the first chapter, when all the remaining polar ice melts, sea level is projected to rise approximately 212 feet (64.5 meters). Adding that to the roughly 400 feet that sea level has risen since the last ice age gives a total vertical range of sea level movement of 612 feet (187 meters). Although we must realize the four hundred is not a precise number, but rather gives an approximate scale to cover the ice age maximum.

Six hundred twelve feet is the height of a tall office building, about 47 floors. Think of the IBM building in New York City, the Space Needle in Seattle, Tower 42 in London, or One Prudential Place in Chicago. Looking up or even imagining the height of these skyscrapers makes the magnitude of sea level rise quite real and stunning.

Visualize a 47-floor elevator as a virtual gauge of sea level change. At the peak of the last ice age, just 20,000 years ago, it was at the "ground floor," due to the vast amount of water locked up in the ice sheets across North America and Eurasia. As the ice melted over some 14,000 years our elevator rose with sea level above the 29th floor, a few feet short of the 30th.

For the last 6,000 years, roughly the period during which human society developed, Earth has been in a narrow band of temperate climates with rather stable polar ice sheets and only tiny, slow movements in sea level. Our virtual elevator, following sea level, has been hanging around the 30th floor, where we are today.

Over the last 100 years the elevator started to rise again and is now eight inches higher. It has started moving upward a little faster. Only a few of the passengers are yet aware that they are even moving. More will become concerned once they realize they are heading up. This reality is just setting in. We are all heading upward, with no way off.

While we may only rise half a floor this century, there is strong evidence that this elevator is now programmed for a much higher floor. On our current warming trend, all the glaciers and ice sheets will eventually melt, bringing us back to a planetary state that has not existed for over 30 million years. If that happens, our sea level elevator is headed to the 47th floor, leaving our coastal world far below the surface.

Metaphors are never perfect, including this one. Nonetheless, it gives a truly accurate portrayal of the actual heights that sea level has risen over the last 20,000 years, and how much it will rise when Greenland and the Antarctic fully melt. With that big picture perspective, let's take a brief journey through the real world of geologic history.

Timescale is Key

Centuries, millions, shmillions —

what's the difference?

Chapter Three
A Billion to a Million Years Ago

To understand the potential for future sea level rise we will start by looking at the past. The first challenge is to get a sense of timescale. Earth is about 4.5 billion years old. To geologists, anything less than 200,000 years is a "brief event." From this perspective, all known human history back to the first cave paintings created 40,000 years ago has been a very brief event, just a "blink" in terms of the planet's history.

To explain sea level rise and avoid the somewhat confusing geologic timeframes (e.g. Cretaceous, Jurassic, Cenozoic, etc.), I like to use four easily distinguished time periods:

- A billion to a million years ago
- A million to 6,000 years ago
- 6,000 years ago to the present (2000 A.D.)
- The present and future (2000-3000 A.D.)

During the last billion years, some eras were much warmer than today, with absolutely no ice on the planet, and some much colder, with most of the planet covered in ice.[4] Viewed in its entirety, the timeframe of a billion to a million years ago was much warmer than the present. At its warmest, the ocean was approximately 18 degrees F (10 degrees C) hotter than today.[5] The polar ice caps did not exist, and sea level was much higher, as shown in Figure 2-1.

A Billion Years Ago: 18 Degrees Warmer

About 65 million years ago, Earth's temperatures plummeted when a huge asteroid slammed into the Gulf of Mexico, changing the atmosphere and causing the mass extinction of an estimated 75 percent of species, including the dinosaurs. That traumatic event ended the Cretaceous period and began a new climactic era, which lasted about 40 million years.[6; 7]

Some 55 million years ago, another dramatic event (known as the Paleocene-Eocene Thermal Maximum) occurred when a huge amount of undersea methane was released, what scientists often refer to as the "methane mega-fart."[8] Scientists debate what triggered this gigantic release of methane, but agree that it caused extremely rapid global warming. Methane is a powerful greenhouse gas and extremely effective at heating the atmosphere. In less than 20,000 years, the average global temperature rose about 11 degrees F (six degrees C). That huge change even made the Arctic rather balmy, with ocean temperatures as warm as 50 to 60 degrees F (10-15 degrees C).[9]

During that mostly warm era hundreds of millions of years ago, sea level was about 400 feet higher than it is today. [130] Recalling the recent explanation that the maximum potential sea level height is approximately 212 feet higher than at present, you might wonder how sea level could have gone up so much higher than it can now when all the ice melts.

The answer has to do with the uplift of continents and the shape of the ocean basins. Uplift occurs from forces such as the earth's plates moving and colliding, in turn forcing up land, similar to the way a sheet of paper lying on a flat surface will bend when you push both ends closer together. Uplift also occurs when ice sheets melt, removing some of their enormous weight and causing the land mass beneath them to rise. This explains how vast amounts of fossilized marine life can be found more than a thousand feet above present sea level in places like Kansas. Even the top of Mt. Everest, the tallest peak on the planet, is made up of marine sediments as a result of uplift.[10]

Present-day San Clemente Island, off the coast of San Diego, shows good evidence of ancient higher sea levels as well as "uplift" of the land forms due to tectonic forces. Those terraces were cut by waves and erosion and mark ancient shorelines, now far above sea level. (Photo, courtesy of the U.S. Geological Survey/Dan Muhs.)

Scientists have found evidence of ancient beaches and erosion points all around the world. Some old coastlines are far underwater; others are now high and dry. San Clemente Island, near San Diego, has a series of distinctive terraces on its hillside, as shown in the image above. Each terrace was created by long-term wave action and represents an ancient shoreline.

The other factor that allowed the oceans to reach greater heights hundreds of millions of years ago is that ocean basins were shaped differently than they are today, again because of the land masses moving around. Smaller areas allowed sea level to achieve greater heights—like pouring the same amount of water into a narrower, taller container.

Thinking about huge changes in sea level often raises questions of where the water came from and whether the quantity of water has remained constant. Many scientists believe that water was formed in the early years of Earth's history, when the planet was intensely hot. Another theory is that large quantities of water arrived on Earth billions of years ago in the form of huge frozen comets.[11]

Whether that is true or not, for the timeframe we are interested in—centuries and millennia—scientific consensus is that the amount of water on Earth has not changed substantially.

What does change is the form that water takes, whether liquid, solid, or vapor. During ice ages, water evaporates from the ocean, then returns to the ground as snow and ice that can last thousands of years on land. As the ice becomes thousands of feet thick, it significantly reduces the amount of water in the ocean. Eventually the ice melts, once again raising sea levels.

Causes Of Cooling And Heating

For nearly three million years, we have had regular ice ages about every 100,000 years. Several factors have caused these cyclical warming and cooling patterns.

The amount of heat Earth receives varies due to the overlap of three orbital cycles: the wobbles of the earth on its axis, the tilt of the axis, and the earth's elliptical orbit around the sun. It turns out that these three cycles, now known as the Milankovitch Cycles, regularly lead us into ice ages every 95,000-125,000 years. Milutin Milankovitch, a Serbian scientist, brilliantly correlated the calculations early last century, but his discovery was poorly recognized until it was finally translated to English in 1969. His discovery largely answered the important geophysical question about what caused the ice ages cycles.

One important phenomenon demonstrated by the Milankovitch Cycles is that subtle temperature changes can have dramatic impacts on our climate. The Milankovitch Cycles cause a variation of much less than one percent in the heat we receive from the sun over the course of a year.[12] Amazingly, that small variation is enough to trigger an ice age, showing just how finely balanced our climate system is. Just that small change causes tipping points and feedback effects that bring on the larger impacts. It's worth noting that the difference of global average temperature between the peak of an ice age and our present climate is less than nine degrees F (five degrees C).[13]

Even the movement of the continents, the plates of the earth's crust, over millions of years, can change climate. For example, when the continents of North and South America joined approximately 3.5 million years ago, they created what is now Central America, blocking the equatorial flow between the Pacific and Atlantic. That connection had a tremendous effect on ocean currents, forcing warmer water up towards the poles, resulting in a new global climate regime. The ice age cycles are one of the signature results emanating from that change in marine circulation patterns.[14; 15]

The *pace* and *extent* of the earth's cooling and heating cycles is also influenced by the development of plant and animal life, particularly in the sea. Plants consume carbon dioxide and produce oxygen, helping to cool the planet; animals do the opposite. As different forms of life have proliferated over tens of millions of years, the composition of the atmosphere—the relative levels of oxygen and carbon dioxide—has changed significantly. But, we will see in upcoming chapters, the three parameters of temperature, carbon dioxide, and sea level always move in synch over the long term. Let's take a look at the evidence.

Chapter Four
A Million to 6,000 Years Ago

In geologic terms, the periods when the earth has large ice sheets are the "ice ages." Thus, technically we are still in an ice age, since there are ice sheets and glaciers in the polar regions. However, in common usage, *ice age* describes the glacial periods when the ice sheets are at their maximum. That is how I will use the term.

With the exception of a "brief" extra high sea level 120,000 years ago, we can characterize the last million years as having substantially lower sea level than at present, due to the vast amount of water locked up in the ice sheets.

The last ice age maximum was just 20,000 years ago—in geologic terms, the blink of an eye. At that time, sea level was almost 400 feet lower than today. Giant sheets of ice, more than two miles thick in some places, extended southward as far as the Mid-Atlantic region of the U.S. and well into Europe and Asia.[16] As the ice melted the ocean rose.

Evidence of these massive ice sheets having removed vast amounts of water from the ocean can be seen in ancient shorelines, in the sedimentary rock record, marine fossils, and in the composition of deep-sea sediments. From such diverse data, a clear picture emerges.

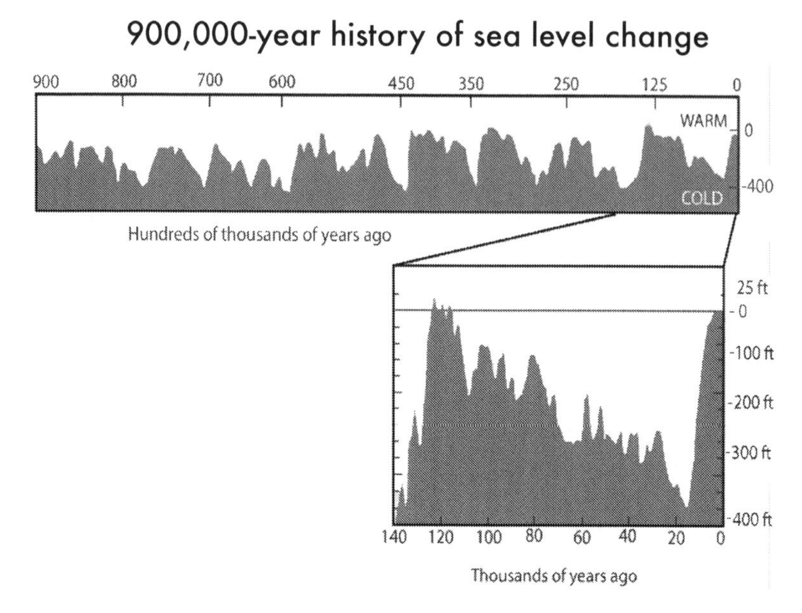

Figure 4-1. 900,000-year history of sea level change, with an expanded view of the most recent 140,000 years. Note the regular cycles of sea level change of up to 390 feet (120 meters), approximately every 100,000 years. The ocean is now roughly at the highest part of this natural cycle of the last few million years. (Graphic, courtesy of NOAA.)

Figure 4-1 shows the record of changing sea level going back 900,000 years, with an enlargement of the most recent 140,000 years. Notice that the ups and downs of sea level occur in what has been a predictable natural cycle. The most recent low point was about 20,000 years ago, and the present sea level height is roughly at the usual high point of these cycles. How do we know the height of these ancient sea levels? There is lots of evidence.

Finding Old Beaches By Submarine

On several occasions I have been on research submarine dives, and have seen ancient shorelines. Most recently, in March 2012, using the Super Aviator submarine, I located evidence of an old coastline off Maui, now 200 feet below the ocean surface. Geologists from the University of Hawaii estimate that this shoreline was created approximately 15,000 years ago. On another occasion, in the Harbor Branch Oceanographic Institute's research submarine *Clelia*, I observed ancient beaches at a depth of about 300 feet. There are many distinctive ancient beaches and shorelines going back to the last ice age that are now underwater at depths ranging to hundreds of feet. Some are quite shallow however, and may be visible to scuba divers or even snorkelers.

Further proof that ancient sea levels were once far below the present levels can be found in underwater caves. If you have ever been in a cave on land, you likely saw the gorgeous stalactites extending down from the ceiling and stalagmites rising up from the cave bottom. These structures grow slowly, as dripping water deposits calcium, which builds up on cave roofs and floors. They cannot grow underwater, as the calcium would instantly dilute in the surrounding water. On advanced cave diving expeditions in the Bahamas, Florida, and in Mexico's Yucatan I have seen vast roomfuls of these stalactites and stalagmites. These rock-like structures have been documented as deep as 300 feet underwater, confirming the extreme range of sea level's downward cycle.

A highly trained scuba diver in an underwater cave, with stalactites above and stalagmites below. These structures are formed by the slow dripping of water that is super-saturated with calcium. They can only form above water. Now a hundred feet underwater, they are one of the many confirmations that sea level was once much lower. Layers within the structures can be radio-carbon dated, accurately showing when they were created. Such techniques greatly enhance the accuracy of our geologic understanding of ancient sea levels. (Photo, courtesy of Stephen Frink.)

CLIMATE DETECTIVES

To explain how we know ancient temperatures and levels of carbon dioxide, it helps to consider another concept with which you are probably familiar: carbon dating. For more than four decades, scientists have been able to determine the approximate age of archeological materials up to 60,000 years old by radio-carbon dating. In simple terms, they look for the amount of the naturally occurring radioactive carbon-14 in a sample, as compared to the non-radioactive form, carbon-12. All living matter contains carbon. Carbon-14 disintegrates at a rate of a specific number of atoms per minute. Scientists studying the age of any object use carbon-14 to calculate the last date that a particular piece of carbon was part of the active biosphere.

Ancient temperatures can be determined using a similar method that measures the isotopes oxygen-16 and oxygen-18. This technique, perfected in the last few decades, uses the relative ratio of these isotopes as a good substitute, or proxy, for temperature, due to their different rates of evaporation. Such temperature records can be found in air bubbles trapped in ancient snow that was compacted and pressed into ice.

When you put these ancient ice chunks in a glass of water, you actually hear hissing and popping sounds as the tiny pressurized bubbles explode with the melting ice. The cross-section of an ice core, pictured in Figure 4-2, shows the micro-bubbles, noticeable as bright spots. The Antarctic (south pole) and Greenland, where seismic data show ice to be the thickest, are the best places to sample ancient ice. Drilling down, scientists have found layers that date back as far as 800,000 years.

Decoding ice layers is similar to using tree rings to determine the age of a tree, whether a year was normal or drought, and if it was exposed to fire or other environmental impacts. In the case of ice layers, we can determine the level of different gases in the atmosphere, find ancient temperature records, and even identify changes in plant life from trapped pollen.

When separate teams began studying ancient temperatures and levels of carbon dioxide using ice cores in very diverse locations in the 1990s, their findings matched, lending strong credibility to the research methods. (Two books about the decoding of the ice cores are *Climate Crash* by John D. Cox, and *The Two-Mile Time Machine* by Richard B. Alley.)

One of thousands of ice cores extracted from ice sheets on Greenland and Antarctica by drilling down thousands of feet. These ice cores contain micro bubbles of ancient atmospheres that indicate temperatures and carbon dioxide levels that can be identified by year. They are stored and studied in special, ultra-cold laboratories to preserve their crystalline structure. (Photo, courtesy of Heidi Roop, National Science Foundation.)

The micro-bubbles of carbon dioxide and the oxygen markers for ancient temperatures produced a very clear view of temperature patterns over 400,000 years of history. As illustrated in Fig 4-2, the parallel patterns clearly show that global temperature and carbon dioxide levels move in unison.

Temperatures from earlier than 800,000 years ago can be deduced from layers of deep ocean sediments that also provide isotope evidence of temperatures and carbon dioxide levels. That kind of research is almost exclusively the province of the massive scientific drill ships that remove deep core samples, which are then carefully studied by dozens of scientific teams internationally.

There is one more intriguing method for observing ancient sea levels, marine archeology, the study of ancient civilizations that are now submerged underwater, some going back more than 10,000 years. Many of these sites, from India, Malta, Japan, and elsewhere, were extensively documented in the fascinating volume, *Underworld*, by Graham Hancock.[17]

CO_2 Concentrations and Proxy Temperatures from the 400,000+ years Vostok Ice Core Data Set

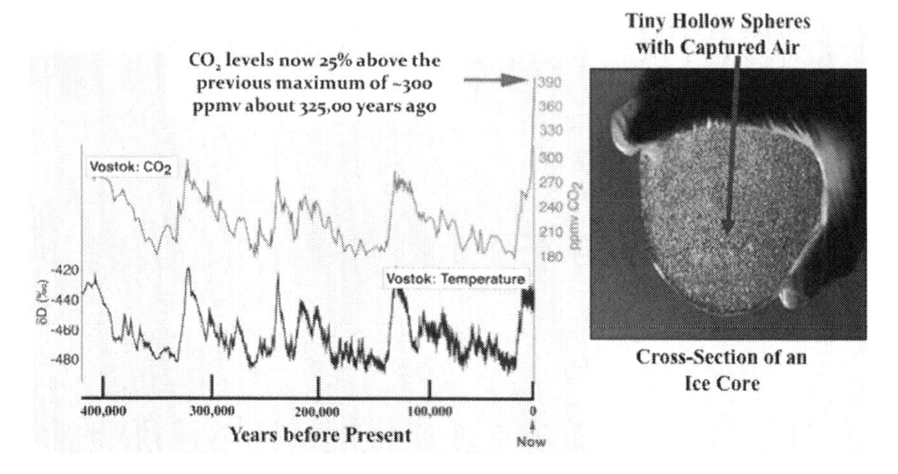

Figure 4-2. On the right, a researcher holds an ice core sample that shows thousands of air bubbles, each containing samples of ancient atmospheres that can be dated to within a few years, with accurate measurements of carbon dioxide levels, as well as isotopes that indicate the temperature levels. The graphs at the left show 400,000 years of deciphered carbon dioxide and temperature. (Graph, courtesy of NOAA.)

ICE AGES AND EXTREME SEA LEVELS

A central question of this book is how fast sea level might rise this century. We can start to uncover the answer by examining prior rates of rise.

As discussed, for the past three million years, the earth has gone through major warming and cooling cycles approximately every 100,000 years. These cycles are triggered by small changes in the amount of heat that reaches the northern latitudes.

Figure 4-3 shows the temperature trends over different time scales. The top graph shows that in the last 65 million years the temperature high point occurred 50 million years ago, when that previously-mentioned tectonic plate jammed into Asia, becoming the Indian subcontinent and causing the rise of the Himalayas. The rise of that massive mountain range, the largest in the world, had a direct impact on global weather, from changes in high altitude air currents to the huge mass of year-round snow and ice at those high elevations. Thus

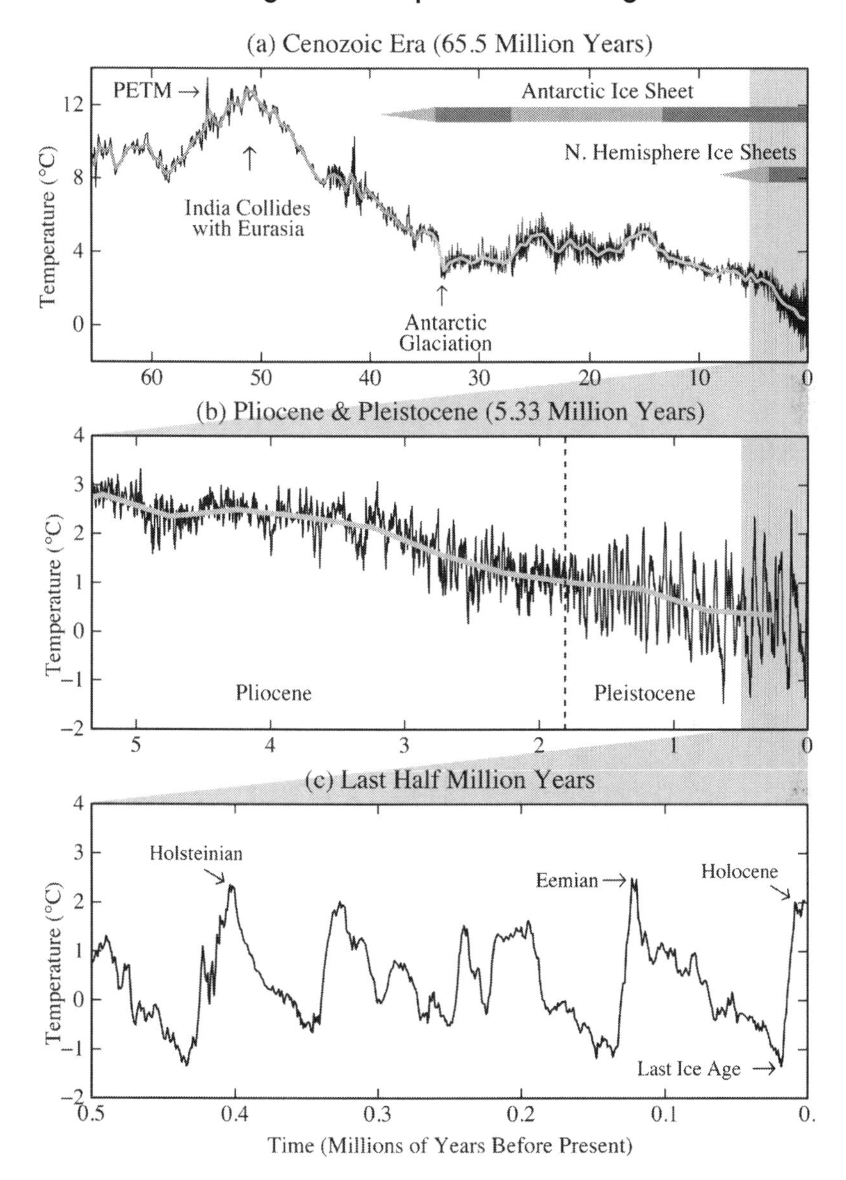

Figure 4-3. The top graph shows the long-term temperature changes for the last 65 million years. The second and third graphs expand the time frame above for greater detail. Note the slow general cooling, since the movement of the continents formed India and the Himalayan mountains, about 50 million years ago. (Graphic, courtesy of Hansen and Sato.)

began a very long-term, gradual cooling trend, with a couple of spikes in temperature along the way.

About three million years ago the up-and-down pattern of the ice ages began. Now we are entering a new era, with temperatures rising beyond the ice-age pattern, taking the planet to a place it has not been in more than 10 million years.

Looking at the last few million years, the cyclical trends of ice ages become very clear. Remember that sea level patterns largely duplicate temperature patterns, reflecting the growth or melting of the polar ice sheets.

The bottom section of Figure 4-3 displays the last 500,000 years, and shows that we are now roughly at the peak of the warming trend for the latest 100,000-year ice age cycle.

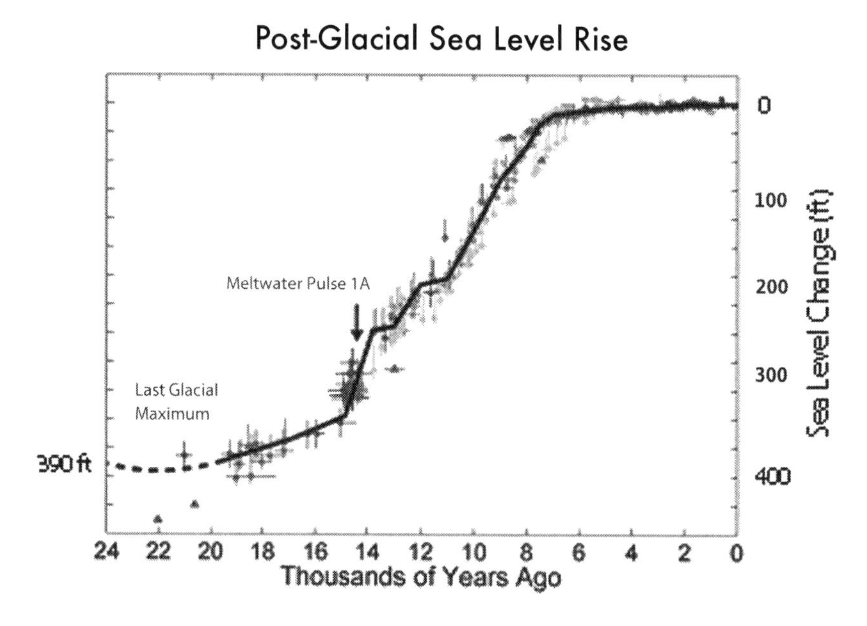

Figure 4-4. This graph shows the ocean was approximately 400 feet lower at the peak of the most recent the ice age, 20,000 years ago. The "Meltwater Pulse 1A" is the period when sea level rose more than a foot per decade for more than four centuries. We are now warming much faster than that natural warming period. (Graphic, by Robert A. Rohde from Wiki Commons.)

This warm era is known as the Holocene, the interglacial period that has existed for about 11,700 years, following the last ice age.

It is worth noting that the warmest point of the previous interglacial (the Eemian) was just a couple of degrees warmer than now, and sea level topped out about 26 feet (eight meters) higher than at present.[18] This occurred with the temperatures being only slightly warmer than today's. On our current path of warming there is no doubt we will shoot past that ocean height. The big question is how high and how soon.

As the ice sheets melted, sea level rose, sometimes gradually and sometimes quite suddenly. For example, 14,000 years ago the sea rose a stunning 65 feet in just four centuries, as much as 18 inches (50 cm) in a single decade. Recent research shows that a major meltdown in Antarctica was associated with this event, known as Meltwater Pulse 1A (see Figure 4-4). Clearly, these dramatic changes in temperature were completely natural variations of the planet's climate.

From that dramatic natural sea level rise following the last ice age, Earth shifted into the present 6,000-8,000-year period of stable climate, stable ice sheets, and stable sea level. During this time, our species and civilization flourished. Most of us just assumed that is how things have always been and always would be. It is important to look at why sea level and climate were stable during these recent few millennia in order to understand why we have almost certainly left that era and entered a new one.

Chapter Five
Past 6,000-8,000 Years

In thinking of sea level history, it's interesting to examine the many flood myths among different cultures. Perhaps the best-known examples are the biblical and Quranic accounts of Noah's Ark, the flood myths of the Quiché and Mayans, the Gilgamesh flood myth and the Hindu puranic flood story of Manu. Though they generally are seen as symbolic and even apocryphal narratives, changing sea levels may well have been a basis for many of these parables. None of those stories has sea level change as a repeating cycle, which makes sense since it takes a perspective of at least 200,000 years to see the long-term patterns of rising and falling sea levels.

There are exceptions to the historical human ignorance of sea level change. For example, the Haida Indians, a prominent native tribe of the Pacific Northwest, have their cultural center on Haida Gwaii, a group of islands off the coast of British Columbia and Alaska. (Until 2010, Haida Gwaii was known as the Queen Charlotte Islands.)

The Haida have lived there continuously for at least 13,000 years.[19] Indeed, the Haida are aware that sea level rises dramatically. Figure 5-1 shows how this puts them on location when sea level rose at an extremely rapid rate of a foot a decade. A fascinating, prominent part of Haida oral history is the mythological story of "The Flood," which describes dramatically changing sea level. As shown in the following translated sections from a modern written version, they even used terms relating to extraordinary tides:[20]

> *...the tide reached low ebb, and the strange woman sat down at the water's edge. The tide began to rise, and the water touched her feet. So she moved up the beach a little and sat down. The water rose up to her, and again she moved back. Before long she reached the high-tide mark, but still the water continued to rise and the tide kept on rising; never before had the tide risen so high.*
>
> *The villagers grew frightened, wondering how they could escape the flood. The chief, very determined to save his fellow tribesmen, told them to gather all the drift logs which were now floating around them, and to tie them strongly together to make a large raft, large enough to carry the whole population.*
>
> *Meanwhile, the stranger continued to sit until the tide reached her. Then she would move to higher ground, above the beach and the village, up the hillside, up to the slopes of the nearby mountain, until at last she sat on its very peak. The whole island now was covered by the sea.*

Haida cultural awareness of large sea level change is unique, as far as I am aware. The fact that anthropologists document the Haida as among the earliest settlers of North America may give them a special vantage point to see such long-term change. A highly disciplined system of teaching oral traditions to the young may also account for the fact that they have retained the story of the rising tide across hundreds of generations.

Having spent a little time with the Haida I find it interesting that these people are far more in touch with the natural realities of our Earth than our modern, fast-paced society that is largely disconnected from it.

Global sea level has barely changed in the last 6,000-8,000 years. It is fascinating and important to understand why global temperatures and sea level were so stable during this long period. Eight thousand years ago, sea level was about seven feet lower than it is today. It rose in two spurts, over 3,000-4,000 years, and reached the current level just over 2,000 years ago. That change was enough to devastate some coastal villages, but gradual enough to allow for human adaptation, particularly for the simple communities of that era.

Temperature during this time was quite stable. Figure 5-1 shows how temperature fluctuated in a rather narrow range during the last 8,000 years. Note the rather small variations for the past 10,000 years, and then the huge warming projected between now and 2100, for comparison.

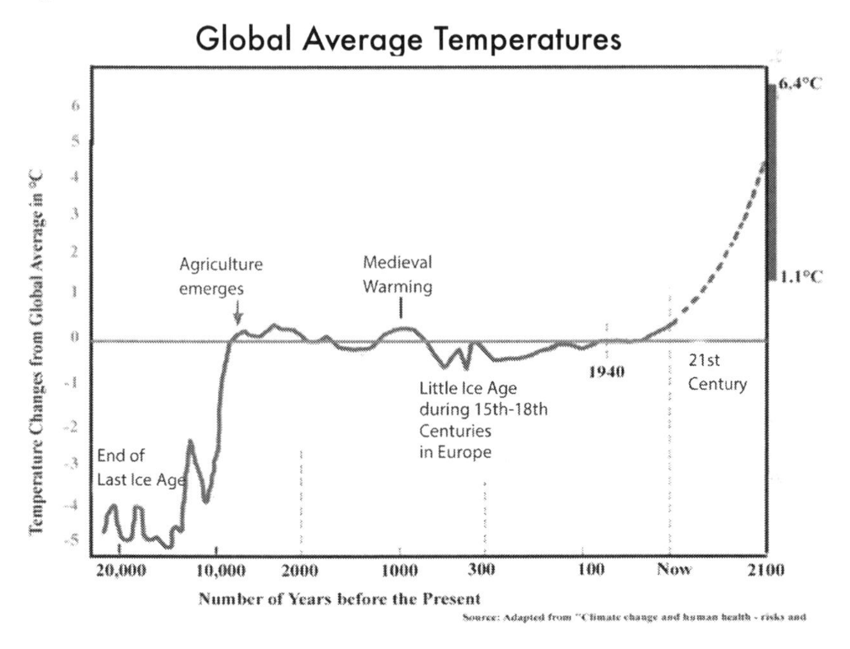

Figure 5-1. Global average temperatures have warmed considerably since the last ice age ended, approximately 20,000 years ago. The small fluctuations in temperature, such as the medieval warm period and the "little ice age," may have been limited to regions such as Europe, and not representative of global averages. Note that the timescale on this is "quasi-log," showing the recent time in greater detail. (Graph courtesy of Climate Change and Human Health: Risks and Responses. *A. J. McMichael, et al. Geneva: World Health Organization, 2003.)*

The 2005 book *Plows, Plagues and Petroleum* by William F. Ruddiman, Ph.D., offers one fascinating new hypothesis about humans' influence on climate stabilization. He suggests that man's impact on climate began with the initiation of agriculture, and that farming helped to stabilize greenhouse gases and temperatures, which would otherwise have begun entering a cooling phase.[21]

Another insight as to why sea level has apparently leveled off for the past several thousand years is related to the fact that the 100,000-year cycle is not equally balanced, half down and half up.

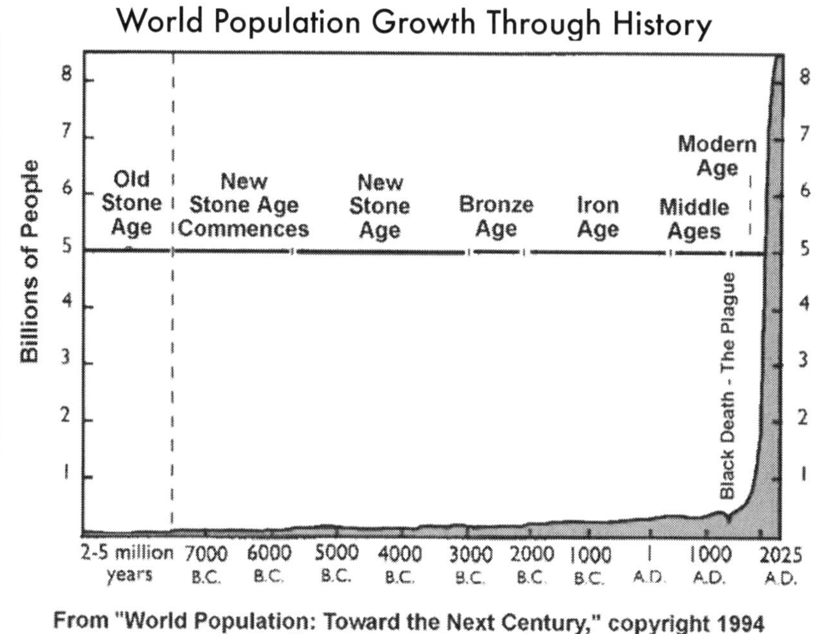

Figure 5-2. Human population levels have changed so dramatically over the last few centuries that it is best to view on a long-scale graph. Just as a point of reference, the "plague" or black death in the 14th century is marked when world population is estimated to have plummeted from about 450 million to as little as 350 million. As shown, this did little to slow the geometric progression. (Graph, courtesy of Population Reference Bureau.)

The ice-building (cooling) phase is more gradual, taking about 80,000 years to complete, followed by a much faster melting (warming) phase that peaks within about 20,000 years. This 80/20 ratio is surprising at first, but makes sense. As temperatures warm over many centuries, the ice sheets melt and the melt water runs into the ocean, a very direct process. The ice-building process, on the other hand, is more complex, and thus much slower. It requires evaporation and then precipitation, which must come down as snow, and then get packed into ice. Thus, common sense explains why it would take much longer to create an ice sheet out of seawater than it does to melt one.

We know that the most recent ice age peaked about 20,000 years ago, and that we are now roughly at what would normally be the warmest point of the 100,000-year ice age cycle. This is corroborated by the fact that sea level is now essentially at the high point of the up-and-down cycles of the last several million years of ice ages as shown in Figure 4-1.

SLACK TIDE

Following that pattern, and assuming no significant human impact, temperature and sea level should be reversing direction and heading into a cooling phase. A lot of scientific evidence suggests that this was actually starting to happen over the course of the last few centuries.[22]

I emphasize *was starting* to happen. Warming and cooling changes in Earth's history are not sharp turns. In fact, the changes in climate can be compared to the *slack tide* in the ocean's tide cycle, when the current changes direction. If you've ever observed fast tidal currents, such as in a coastal inlet, you likely have seen slack tide, when the water is swirling and churning, and it's hard to tell whether the tide is coming in or going out.

The same type of thing happens during changing climate cycles. Turning such a giant complex mass of ocean, vegetation, animal life, and atmosphere from the warming mode into the cooling mode is a slow and bumpy process, with some confusing forces and indications during the turn.

As recently as the 1930s there were signs of expanding glaciers, as would be expected if we were entering the usual cooling period. The stable sea level of the last 6,000 years had been in the midst of the slack tide effect, where it appeared to be stationary, but was actually reversing direction from rising to falling. But that happened to coincide with the explosion of population (see Figure 5-2), technology, and industry that has hugely affected the planet's natural systems and has dramatically shifted us back into an abnormal warming trend that will cause sea level to rise for many centuries to come.

A helpful metaphor may be the classic "tug-of-war" contest where two opposing sides pull on a rope. With strong, complex opposing forces each side can briefly lose ground. As described in this chapter, the natural 100,000-year ice age cycles were about to put earth into a cooling phase, at the same time that population, technology, and energy production were producing an opposing warming force. The "judges" in this contest are the size of the ice sheets, glaciers and sea level.

Chapter Six
The Present and Future (2000-3000 A.D.)

It was only in the 1830's that the concept of the "ice age" was put together by legendary geologist Louis Agassiz with Karl Friedrich Schimper. Even with that, it took almost half a century for the idea to be accepted as a fundamental principle of geology. Today, there is no credible earth scientist who doubts the ice age cycles. It is but one example of how it can take time for the scientific community to fully accept a new concept.

While geoscientists have understood the connection of ice ages and sea level rise for roughly a century, the concept is still largely unknown among the general population. Call it naïve, ignorant, or unenlightened, around the world we have erected trillions of dollars worth of commercial buildings, homes, and infrastructure on the coast, based on the illusion that the sea will not rise, causing the shoreline to move.

That was not a practical problem until the recent rapid warming began, commencing a new era of rising ocean. In the twentieth century, sea level rose about 7.5 inches; in the last three decades the rate of increase doubled.

Changes in sea level from year to year occur for several reasons. Most important is annual temperature change, specifically, whether the temperature rises enough to melt glaciers and other snowpack. If the temperature is near freezing, just a single degree of warming or cooling will determine whether that frozen water remains on land, or melts and finds its way to the ocean, raising sea level. Glaciers and snowpack cover vast areas of the planet. Particularly at the boundary areas, just a small temperature change will greatly affect whether melting occurs. It takes a lot of melting to affect global sea level, but we have a lot of snow and ice to melt, and that is now happening.

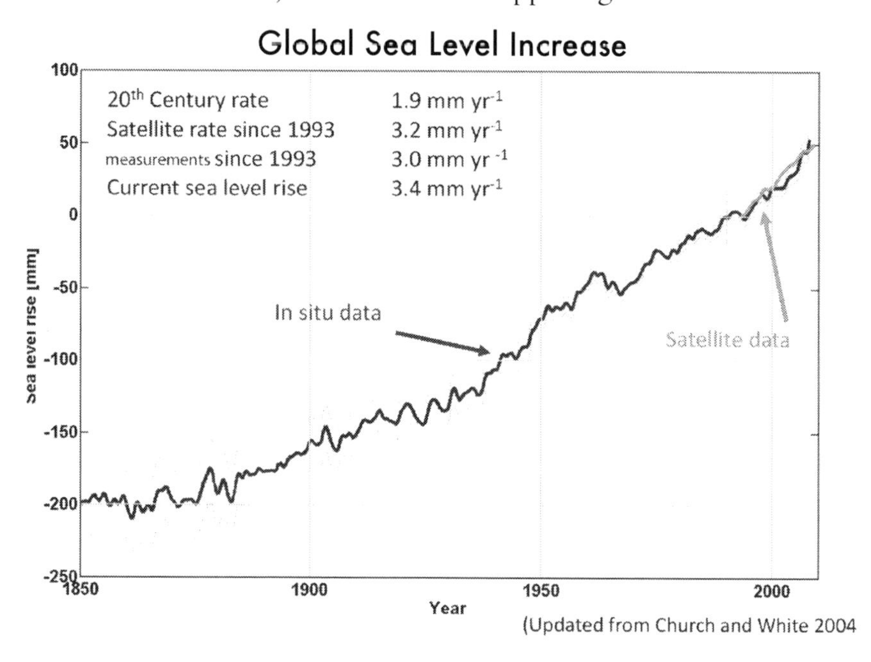

Figure 6-1. From 1900-2000, global sea level increased by about 7.5 inches (20 centimeters), an average annual rate of 1.9 millimeters per year. Very significant is the fact that in the last three decades, the rate of rise has almost doubled to 3.4 millimeters. (Reprinted from Global and Planetary Change, *Vol 72/Issue 3, A. Webb and P. Kench, "The dynamic response of reef islands to sea-level rise: Evidence from multi-decadal analysis of island change in the Central Pacific," p. 234-246, 2010, with permission from Elsevier.)*

Annual changes in rain patterns—normal versus droughts and floods—also determine whether the soil absorbs water or runs off and quickly finds its way to the ocean. In addition, the total amount of water stored by dams from year to year can affect very short-term sea level changes. All of these variables mean that within a decade that has a clear warming, there can still be cooler years or warmer years with more cloud cover, which slows the melting of the icecaps. It's complex, convoluted, and provides lots of material for anyone wanting to confuse the issue.

Climatologists continuously strive to understand all the nuances of climate change. They study it and model it, similar to the way experts study, model, and predict a flu epidemic, or the economic world. The models improve every few years, and are continually refined.

65 Feet Of Sea Level Rise Per Degree

The issue of climate change is now probably the single greatest concern of the overall scientific community, and sea level rise is one of the most profound consequences of climate change. Thousands of scientists worldwide are working to understand its potential impact.

One way that scientists predict future scenarios is by looking at the geologic past. In an important 2009 article in the journal *Science*, lead author Dr. Aradhna Tripati wrote:

> *Fifteen million years ago was the last time carbon dioxide levels were apparently as high as they are today, and were sustained at those levels. Global temperatures were five to ten degrees F higher than they are today (2.5 - five degrees C); the sea level was approximately 75 - 120 feet higher than today; there was no permanent sea ice cap in the Arctic and very little ice on Antarctica and Greenland.*[23]

This may be the single most important piece of information in this book, so I will restate. This scientist's highly-qualified team published a study, rigorously reviewed by peers, which found that it has been 15 million years since carbon dioxide levels were as high as they are today. Such high levels of carbon dioxide corresponded with global temperature about 10 degrees F (six degrees C) hotter and sea level roughly 100 feet (30 meters) higher than today's.

In 2008, Dr. David Archer showed what happens over longer periods of time when temperature, ice sheets and sea levels reach a new equilibrium. Using the work of glaciologist Dr. Richard Alley, his analysis showed that sea level has changed by a stunning 20 meters (65 feet) for every degree of change Celsius (1.8 degrees F). Since the oceans and atmosphere have already warmed almost one full degree C, it is just a matter of time before sea level adjusts according to that historic relationship.

Think about that. Based on 40 million years of actual history, not theoretical projections, sea level will rise more than 50 feet from the 0.8 degrees C of warming that has already happened, once the ice has had sufficient time to melt, however long that may be. Limiting that rise to 50 feet assumes that there is no further warming, essentially an impossibility. Many international groups are desperately trying to figure out how to keep the temperature this century from rising more than an additional 3.6 degrees F (two degrees C).

This Century: Uncertain And Ominous

Scientific teams from the U.S., the U.K., Japan, Germany, China and many other countries are working collaboratively to create sea level rise projections using supercomputer models. Even using the fastest supercomputers, the climate models are so complicated that it typically takes more than a week to run a single set of calculations.

The Climate Scoreboard (climatescoreboard.org) has created an elegant, simple thermometer to help visualize climate model changes (see Figure 6-2). Three numbers are shown for the year 2100. "Business as usual" is the projection of global mean temperature increase, assuming we generate our power from the usual fuel sources and that energy demands continue to grow with population and international development. "Proposals" represents the latest international negotiations and agreements to reduce greenhouse gas emissions this century. "Goals" is the level of temperature increase that governments and scientists say is the upper limit of temperature increase to avoid dramatic, possibly catastrophic, changes to our climate.

As shown, unless something changes soon in our policies and practices, we will have seven to eight degrees F (four to five degrees C) of warming by 2100, which will melt the ice sheets significantly, and lead to dramatically higher seas.

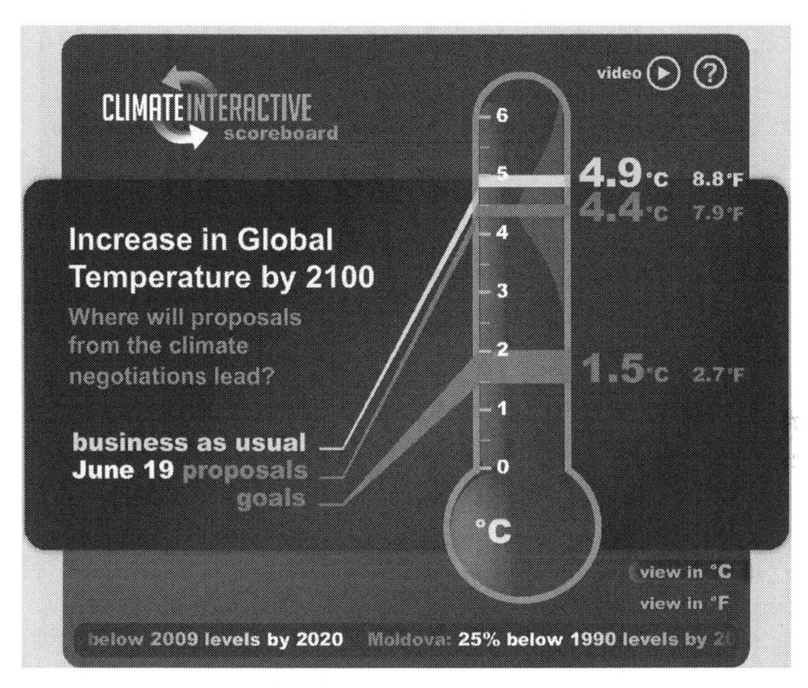

Figure 6-2. Climate Interactive maintains a simple thermometer graphic depicting what the supercomputer models show for temperature projections this century. "Business as usual" represents our current path of emissions. The most recent international proposals show a slight reduction. The bottom figure is the "goal" that has been set to maintain a relatively sustainable climate similar to the last few thousand years. Visit their website to see the latest update. (Graph, courtesy of Climate Interactive, climatescoreboard.org/scoreboard.)

Figure 6-3 depicts several sea level rise estimates for the year 2100, from a low of less than one foot, up to a seven foot rise. Predictions vary based on assumptions made about Earth's population, total energy consumed each year, energy sources, and the corresponding carbon dioxide emissions. Uncertainty comes from the fact that the rate of warming this century is without precedent in the known geologic record.

You might recall that in chapter three we looked at catastrophic natural climate change such as the PETN event 55 million years ago when global average temperature rose about eleven degrees in 20,000 years. The projected warming now will almost equal that increase this century, 200 times faster.

It is this unprecedented *rate* of warming that is cause for huge concern. Many scientists believe that five feet is the maximum amount that sea level could rise this century,[24] while at least one very credible expert says that it could be at least double that figure.[25]

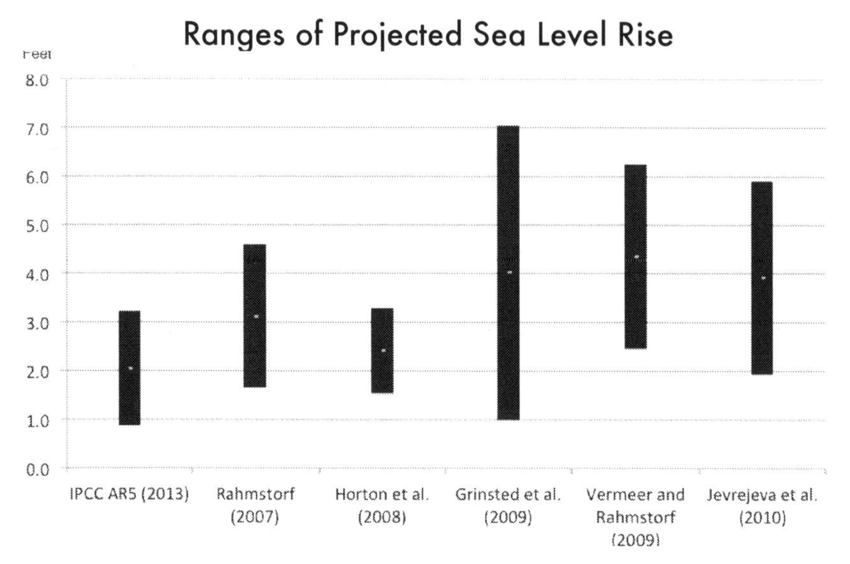

Figure 6-3. Six different projections for sea level rise by year 2100 range from 10 inches (26 cm) to 7 feet (2 m). Large variation results from varied assumptions about the rate of greenhouse gas emissions and sources of energy over the course of this century. IPCC figures did not include accelerating melt rates in Greenland or Antarctica. Projections do not reflect the possibility of catastrophic melting. (Data from Rahmstorf, 2010 updated for IPCC 2013.)

TIPPING POINTS

One reason for the range of projections for sea level rise is differing assumptions about tipping points, when patterns change abruptly, dramatically, and sometimes permanently.

An easy-to-visualize example of a tipping point is snow on a roof. If snow falls on the roof at a constant rate, and accumulates until the weight exceeds the structural capacity of the roof, it will suddenly sag or collapse. Just by watching the snow accumulate you would never know what the tipping point is. You can only determine it with certainty by running an experiment to calculate when the trend will shift abruptly.

Sea level rise is subject to several tipping points. An important one relates to the temperature at which ice melts. Whereas temperatures can warm for a very long time without significant disruptive change, at the magic temperature of 32 degrees F (zero C) everything changes dramatically. Solid ice becomes liquid, which turns the white, reflective surface of ice sheets into nearly black, heat-absorbing ocean, which causes the planet to get warmer, which then melts more ice, creating what is known as a positive feedback loop. We explain this effect more later.

Another, equally powerful tipping point is hidden away in the molecular bonds of ice. By definition, it takes one calorie to heat a tiny cubic centimeter of water one degree C. However, if that tiny cubic centimeter is ice, it takes 80 calories to heat it one degree, breaking the molecular bonds and turning it into water. So melting ice consumes a lot of energy. Once the ice is melted that same amount of heat will significantly increase the water temperature. As a result, as the ice disappears, the warming dramatically speeds up, melting more ice. The shrinking Arctic icepack accelerates the warming. Whether you think of it as a positive feedback loop, or a tipping point, it is a key reason that the warming and ice melting are speeding up so noticeably.

The sudden rise in sea level 14,000 years ago is no doubt an effect of a tipping point, in which large areas of ice were exposed to melting temperatures, causing things to change dramatically.

The sea level rise forecasts described so far do not include the possibility of so-called catastrophic melting,[26] which would essentially be sea level tipping points in action.

At least one nation, the Netherlands, is considering such a scenario. Much of that country is below sea level, making it very vulnerable, but giving it relevant experience. The last major flood, in 1953, killed 1,835 people overnight, leaving permanent scars in the landscape and the culture.

In response, over the next four decades they built the Delta Works project, a massive system of defense against high water. The country now views four feet (1.3 meters) of sea level rise this century as a minimum projection and is considering much more extreme scenarios.[27]

For more than 30 years, one of the world's leading thinkers about climate change has been National Aeronautics and Space Administration (NASA) scientist, Dr. James Hansen. An astrophysicist, originally he studied the atmosphere of Venus.

In 1978 he turned his attention and research to Earth, noting that our atmosphere was changing quickly, particularly the level of carbon dioxide. He recalls thinking, "It would be more useful and interesting to try to help understand how the climate on our own planet will change." It is hard to imagine a greater understatement in the entire history of science, even though he likely was unaware at the time.

He could not possibly have known how much his change in focus would affect his own life, and the scientific understandings relating to our collective future. I had the privilege of spending a little time with him not long ago, where we discussed his greatest concern: that greenhouse gases are pushing our climate past a tipping point that will gravely change the world in which his grandchildren will live.

HANSEN'S NEW PROJECTION

In 2011 Hansen published a paper with an alarming new hypothesis that sent shock waves through the scientific community. He and colleague Makiko Sato suggested that our current climate is almost at a tipping point. Based on the historical record and our trajectory of temperature and carbon dioxide, they say that the melting rate in Greenland and the Antarctic could increase in a non-linear manner, *possibly* even doubling every decade. Such a progression would mean as much as 16 feet (five meters) of sea level rise by the end of the century, although 80 percent of the rise would happen in the last two decades. They clearly state that the possible doubling is not a precise figure, but use it to show that the rate of change is not linear.

If the doubling proves correct, that would mean almost a foot of rise per year by the end of the century, and a further doubling of the rising ocean height each decade into the next century, until there was no more ice to melt. At anything like that rate of increase the ice sheets would be gone in a matter of centuries, not millennia, as most of the scientific community has estimated. There is precedent, as was shown in Figure 5-1, of very rapid sea level rise about 14,000 years ago. And today, we are warming much more quickly.

In another decade, we should know whether Hansen and Sato's new theory is accurate. However, by that time it is unlikely that we will be able to claw our way back from the world of runaway melting and sea level rise.

At this point, I should probably go on record to state my own conclusions for sea level rise this century, based on the thousands of hours of research, discussions with leading scientists, and personal observations in the field to corroborate what I have read and heard. I expect we will have sea level rise on the order of one foot or more by 2050. I think a four- to five-foot rise by 2100 is realistic, though that could easily shift upwards with new information in the years ahead.

TOWNS ALREADY UNDERWATER

Though sea level has only started to rise significantly in the last century or so, some coastal communities are already starting to go underwater. Sharps Island off the Maryland coast is one example that you can visit, but you'll need to bring your dive mask and snorkel.

Sharps Island Lighthouse in Chesapeake Bay, not far from Washington, D.C., is now badly in decay and has been decommissioned. Originally, it was on a thriving island, with farms and a hotel that operated until 1910. The island disappeared in 1962, due to the combined effects of rising sea level, erosion, and subsidence leaving just the remains of the lighthouse. (Photo, courtesy of Constantine M. Frangos.)

A few hundred years ago, the 900-acre Sharps Island hosted a thriving community, including a prominent hotel and several large farms. But over time, rising sea level, erosion, and land subsidence ate away at the island. The hotel operated until 1910, just a century ago. Sharps Island finally disappeared into Chesapeake Bay in 1962.

Nearby, on Holland Island in the Chesapeake Bay, the last house disappeared into the surrounding waters on October 21, 2010, having just been abandoned by the owner the previous year. Similar examples exist around the world.

Erosion and sea level rise are not noticeable most days. What commands our attention is when the ocean makes its way into our community and onto main street. This can happen gradually, as occurred in these island communities, or dramatically during extreme tides and storms.

The last house on Holland Island in the Chesapeake Bay as it stood in October 2009. It fell into the bay a year later. The owners held on until almost the last possible minute. (Source: flikr.com baldeaglebluff.)

Timescale Is Key

When we talk about what will happen next century, it might be easy to think, "What do I care?" Well, there may be a very important reason to care: coastal property values.

We must consider the near certainty that most coastal property will be affected, reducing its value. At some point, people will become reluctant to invest in real estate and infrastructure that cannot be resold, transferred or bequeathed because its future is uncertain.

With all of this, timescale is key. The difference between decades, centuries, and millennia can be enormous.

Columnist George Will wrote a ridiculous column in *Newsweek* on September 20, 2010. Without question, he is a true intellectual and has written some brilliant pieces. In my opinion, his *Newsweek* column "The earth Doesn't Care" was an exception.[28]

In it he features the position regarding climate change of Robert B. Laughlin, co-winner of a 1998 Nobel Prize. Will uses his column to showcase Laughlin's attitude of fatalism about man's impact on the climate. Laughlin acknowledges that humans are changing the earth's climate and chemistry, but concludes that in thousands of years, the earth will compensate and rebalance things, making our impact meaningless.

Saying that the planet will not care in 1,000 years or 100,000 years is tantamount to saying that millions and billions of years from now, the expanding sun will obliterate Earth, so why should we bother about anything? While the 70-year-old Will may prefer to take the millennial view, I wonder what his four children or their children will be thinking in the latter part of this century about such a cavalier attitude.

Fatalism and gallows humor about sea level rise are certainly understandable. I firmly believe, however, that the potential, even likely disaster can be dramatically improved with a proactive attitude. First, let's take a look at the forces behind sea level rise and dispel some common myths and misunderstandings about why it is happening.

Fact v. Fiction

Facts are stubborn things.
~President John Adams

Everyone is entitled to his own opinions,
but not to his own facts.
~Sen. Daniel Patrick Moynihan

Then you will know the truth,
and the truth will set you free.
~The Bible, John 8:32

Chapter Seven
On Thinning Ice

You have likely seen satellite images showing that the Arctic ice sheet is shrinking. Perhaps you have heard the forecasts that the Arctic will be ice-free within a decade or two, opening seasonal short-cut shipping routes such as the fabled Northwest Passage. Like most of the issues about sea level rise, you probably have not been told the really important truths.

Most people think that the melting Arctic ice cap is causing sea level to rise. Strange as it may seem, it does NOT affect sea level directly. When you realize that the frozen Arctic Ocean around the North Pole is floating sea ice, it makes sense that it is different than the glaciers on land that melt and flow into the sea, raising its level.

At the same time, an ice-free Arctic is extremely important because the consequences of its melting may *trigger* catastrophic warming. That could very well melt the land-based ice, which is what will eventually raise sea level 212 feet. We need to understand these and other facts to appreciate what contributes to sea level rise and what is irrelevant.

Thermal Expansion Of Seawater

Let's start with the easily missed factor that has actually caused the most sea level rise this past century: thermal expansion. Like most substances, as seawater warms it expands very slightly. Since the average ocean depth is over two miles (three kilometers), even a minuscule amount of expansion translates to measurable sea level rise. Sophisticated satellite technology now allows scientists to identify the annual rise in sea level within tenths of a millimeter and to attribute the increase between thermal expansion and the other factors covered just below.[29] Such precision is necessary in order to achieve an accurate projection over the course of the coming century. Reconciling the diverse data in the hundredths of a millimeter is challenging work, but is being addressed by teams of scientists internationally. (For those interested in the scientific detail, see *Understanding Sea-Level Rise and Variability*, by Church, Woodworth, Aarup, and Wilson, 2010.)[30]

To summarize the outlook for thermal expansion for our purpose of seeing the big picture, on our current path of warming, it is projected to raise sea levels by as much as one foot (0.1-0.3 meters) this century.[31] While not insignificant, the amount of sea level rise from thermal expansion in the longer term will pale in comparison to that from the melting of Greenland and the Antarctic.

Greenland

The Greenland Ice Sheet is huge, covering approximately 80 percent of the island, which is more than three times the size of Texas. Note that I referred to "ice sheet" and not glacier. It is essential to distinguish the two. We have just two great ice sheets on Earth, the huge flat expanses on top of Greenland and Antarctica that contain about 99 percent of the ice on the planet. There are tens of thousands of glaciers on almost every continent. A glacier is a slow moving river of ice. Numerous glaciers gradually move ice from the ice sheets to the sea, often breaking off into icebergs.

In 2007, I organized a five day fact-finding trip to Greenland with the International SeaKeepers Society to see the sudden melting firsthand and to meet with scientists. Immediately upon landing we got some news that made the rapidly changing situation quite clear. We had arranged to go dog sledding as a recreational activity. It was quite a disappointment that the outing had to be cancelled because the melting snow made it dangerous for us and for the dogs. That level of ice melting had never happened before, at least not in five generations of memory in the community of Ilulissat.

The president of the Inuit Circumpolar Council, Aqqaluk Lynge, told me, "As far back as our people can remember, there has never been melting like that of the last decade." That's quite a statement considering that these people have lived all over the polar region for thousands of years, and have a rich oral history.

Currently, Greenland is approaching runaway melt mode. Even as I write this, new reports continue to hit mainstream press describing accelerating massive ice melt. In July, 2012, scientists were alarmed to find that nearly all of Greenland was melting during the melting season, an estimated 97 percent of the island. Up until last year, the melting had not reached the higher elevations.

Figure 7-1. The areas in white mark the sections of the Greenland ice sheet that have melted compared to the previous year. Note the large increase in melt area from 1992 on the left, to 2002, on the right. By the summer of 2012, the melt area had increased to 97 percent of the surface. (©2004, ACIA/ Map ©Clifford Grabhorn.)

That massive ice sheet is as much as two miles (three kilometers) thick in some places, but is losing as much as 100 feet of thickness a year. Figure 7-1 comes from the Arctic Climate Impact Assessment, the consortium of eight nations with territory inside the Arctic Circle. This image shows how the areas of net annual melt have significantly increased in just 10 years, from 1992 to 2002. The updated 2012 version, (not shown here) shows virtually the entire island in a melting state.

The complete melting of the Greenland Ice Sheet will raise sea level about 24 feet. Fortunately, no scientists predict a total meltdown this century, and most say a complete melting will take many centuries, if not several millennia. But even melting a small portion of the total amount will have devastating results.

Geoglacial Features of a Moulin

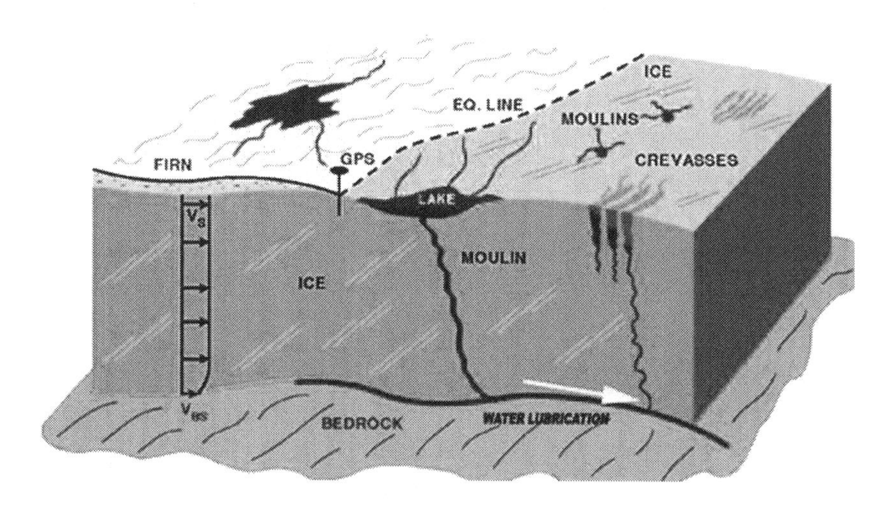

Figure 7-2. The two squiggly lines towards the right illustrate how the melting water at the surface gets down through the several miles of ice, forming a water layer between the ice sheets and glaciers. This layer of water dramatically speeds up the flow rate of the glaciers going towards the sea, directly adding to sea level rise. (Courtesy of NASA.)

In Greenland, the melting is made worse by the way the icemelt drains. The meltwater cascades down more than 100 vertical shafts, called moulins, in huge torrents (see Figure 7-2). The water ends up as a layer beneath the ice but above the bedrock. That stream lubricates the glaciers, making them move much faster than if solid ice were

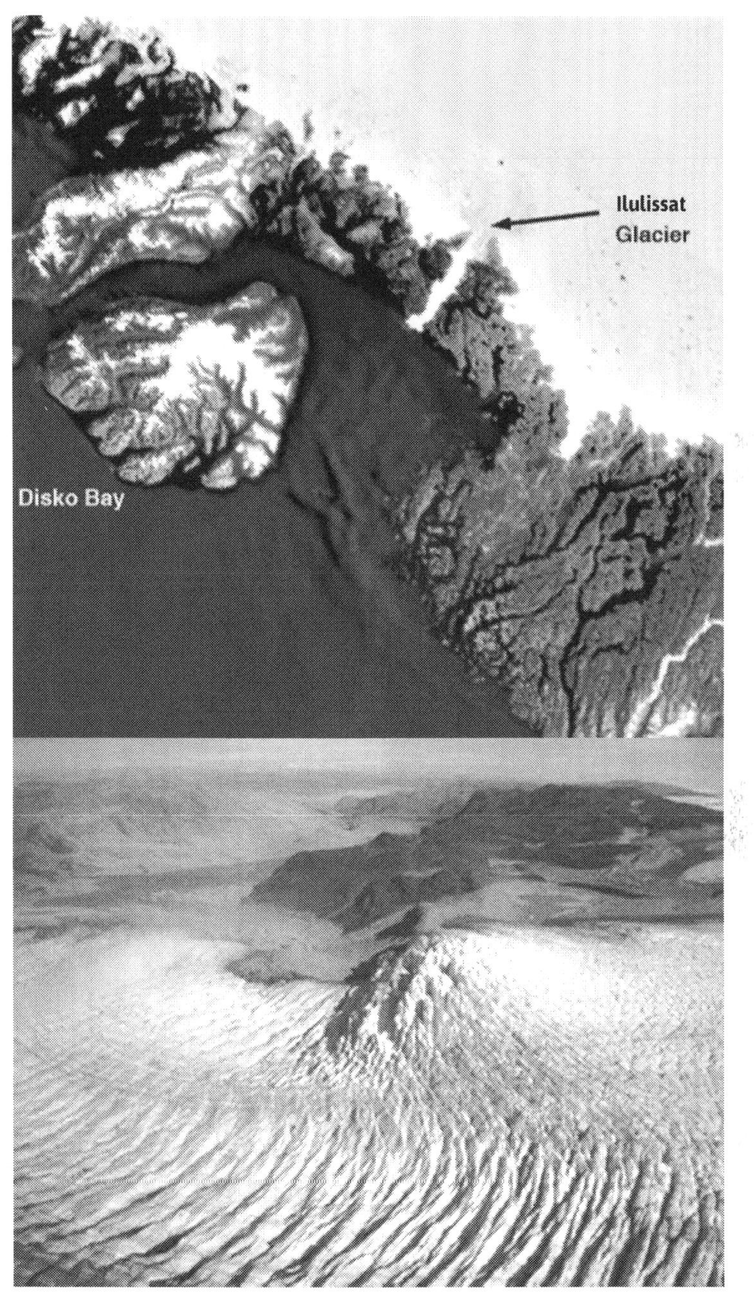

Satellite and aerial views of glaciers in Greenland demonstrate how these massive ice sheets drain to the ocean. In the top photo of Jakobshavn (Ilulissat), the flat, solid white is the ice sheet. A glacier is a river of ice that drains the ice sheet to the ocean as marked on the upper image, and shown in close-up in the lower image of Equi Glacier.

grinding against the rock. The largest glacier, Jakobshavn (also known by Ilulissat - its Greenlandic name), moved towards the sea at a rate of nearly four miles per year in 1992. The rate had fully doubled in just one decade.[32]

The extremely rapid, non-linear increase in melting over the last decade makes it very difficult to predict future glacier melt. It is simply hard to plot the graph given the severe upward increase. Yet, the cumulative change in melt rates over a few decades will make a real difference to global sea levels, and thus to our assets, infrastructure, and lives.

Many teams of scientists from different countries are working to refine the forecasts, using advanced satellites to measure the ice mass, and highly sensitive instruments on the ice to record changes from year to year. In the next few years we should be much better equipped to plot the projected melt rates later this century.

Potential Sea Level Rise – Total Melt of Ice Sheet and Glaciers

Location	Feet	Meters
East Antarctic Ice Sheet	169	51.6
West Antarctic Ice Sheet	15	4.5
Antarctic peninsula	2	0.5
Greenland Ice Sheet	24	7.3
Other glaciers, etc	2	0.6
Total Ice from melting	212	64.5

(Figure 7-3. Adapted from Allison, et. al., 2009.)

ANTARCTICA: THE 185-FOOT QUESTION

In terms of sea level rise, the southern continent of Antarctica represents an even greater unknown than Greenland. The Antarctic is roughly the size of the United States and Mexico combined, and holds almost ten times the amount of ice as Greenland. The Antarctic ice sheet is, on average, more than a mile thick, but can be as much as three miles (five kilometers) thick.

Melting of Antarctica will raise sea level 185 feet. While no scientists forecast the Antarctic ice sheet to completely melt within the next couple of centuries, there are scenarios that could lead to nasty consequences even in the decades ahead. To understand them, we need to have a cursory awareness of Antarctica.

Within this pure-white landscape there are four easily identified regions: the Antarctic Peninsula, the East Antarctic Ice Sheet, the West Antarctic Ice Sheet and the ice shelves (see Figure 7-4).

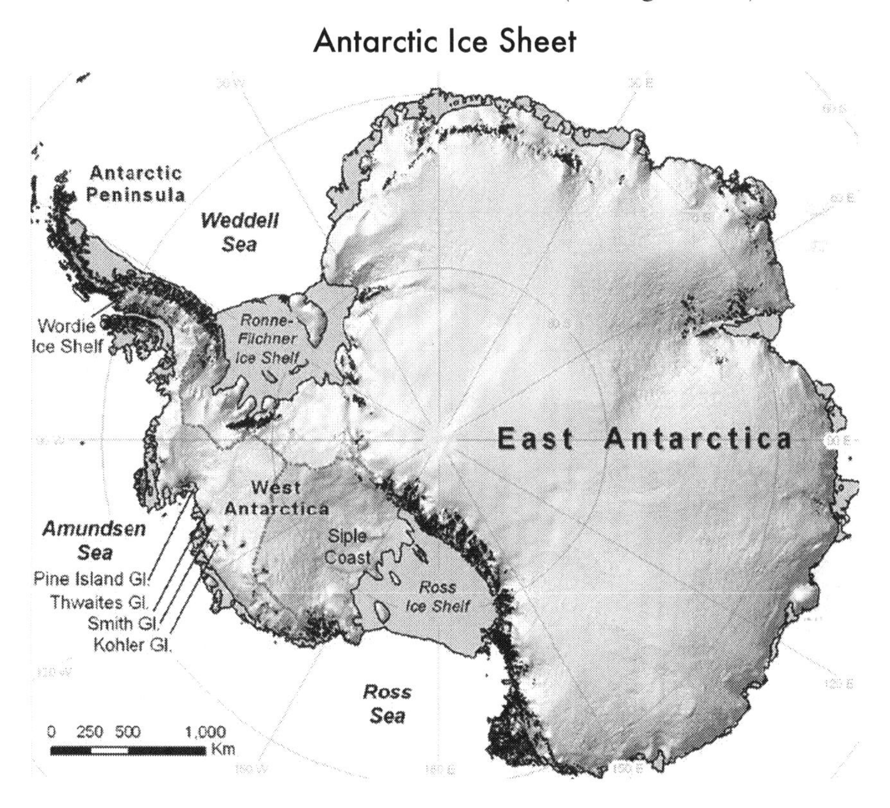

Figure 7-4. The frozen continent has four major distinct areas. East Antarctica, the largest, is still quite frozen and growing as a result of more moist air from the warming ocean, coming down as snow. West Antarctica is showing signs of thawing and could lead to catastrophic sea level rise if its massive glaciers become "uncorked." The fringing ice shelves act as a stopper for the glaciers, but have been dramatically melting and collapsing in recent years. The small Antarctic Peninsula has experienced more warming and melting in the last half century than anywhere else on the planet, likely signaling what will happen for the areas a little further south. (With kind permission from Springer Science+Business Media: Climatic Change, West Antarctic Ice Sheet collapse – the fall and rise of a paradigm, vol 91/Iss 1, 2008, p. 65-79, D. Vaughan.)

The appendix-shaped **Antarctic Peninsula** is warming more quickly than anywhere on the entire planet. In the last 50 years, the mean annual temperatures of the Antarctic Peninsula rose by about five degrees F (2.8 C). This is unprecedented in the last 1,000 years, and likely in the last 10,000 years.[33] Though the Antarctic Peninsula is only a fraction of the total, it may be an early indicator for melting on the rest of the frozen continent.

The **East Antarctic Ice Sheet** is the most stable area of Antarctica. It does not show imminent large-scale melting, and in fact it is gaining thickness in many parts, as will be explained.

The **West Antarctic Ice Sheet** is particularly vulnerable to melting. Whereas the East Antarctic Ice Sheet is a vast sheet on top of level ground, the West Antarctic Ice Sheet is anchored by two large mountains, with a large portion of the ice sheet going below sea level and resting on the bedrock underwater. That vulnerability, plus its massive size, is a recipe for truly epic ice sheet collapse. Dr. John Mercer, the late pioneering Antarctic expert, stated as early as 1968 that the West Antarctic Ice sheet was the harbinger of the "big melt." Changes to the configuration of this sheet could destabilize it, bringing even more ice from the land to the ocean.

The nearly **50 ice shelves** of the Antarctic are located on the perimeter of the continent. Ice shelves form when melting glaciers reach the sea. If the ocean is cold enough, that newly arrived ice doesn't melt right away. Instead it may float on the surface and grow larger as glacial ice behind it continues to flow into the sea. In some cases the ice shelves are solid enough to be supported from the sides, and melt from below, causing them to be suspended over the sea. Ice shelves can remain for thousands of years, holding back huge glaciers like stoppers in bottles. But if the ice shelves disintegrate, uncorking the bottle, the glaciers will slowly pour off the land into the ocean, raising sea level.

In the last decade thousands of square miles of ice shelf have been lost. In some cases, they have collapsed in hours even as people watched on television.[34] Notable was the nearly instantaneous disintegration of a portion of the Larsen B Ice Shelf in early March 2002; it was the size of Rhode Island and more than 650 feet (200 meters) thick.

In an article published in 1978 the iconoclastic Dr. Mercer wrote:

If present trends in fossil fuel consumption continue, and if the greenhouse warming effect of the resultant increasing atmospheric carbon dioxide is as great as the most advanced current models suggest, a critical level of warmth will have been passed in high southern latitudes 50 years from now, and deglaciation (reduction) of West Antarctica will be imminent or in progress. Deglaciation would probably be rapid once it had started, and when complete would have led to a rise in sea level of about five meters (16 feet) along most coasts.[35]

Note, it was 35 years ago when Mercer made that 50-year forecast. If he was correct, the disgorging of the Pine Island and Thwaites glaciers in Western Antarctica would herald catastrophic sea level rise. Recent measurements make Mercer appear to be truly prophetic. In 2006, an article by Dr. Eric Rignot, in the prestigious scientific journal of Britain's Royal Society, stated:

The melting of Pine Island Glacier accelerated 38 percent since 1975, and most of the speed up took place over the last decade. Its neighbour Thwaites Glacier is widening up and may double its width when its weakened eastern ice shelf breaks up. [36]

Another peer-reviewed article in *Science* in 2009 reassessed the probability and magnitude of catastrophic collapse of the West Antarctic Ice Sheet and concluded that a realistic worst case scenario from this area was approximately 10 feet (three meters) of sea level rise this century.[37]

Other scholarly publications echo Mercer's forecast, but caution that it is too early to conclude with certainty that catastrophic collapse is imminent.[38] Considering what is at risk, this seems like a very weak reason to relax and assume that this will not turn into catastrophe this century. To paraphrase a common metaphor: if we are not yet on thin ice, the ice is a lot thinner.

In this second edition of this book (2013) we have a relevant update. In early July this year, a massive iceberg, larger than Chicago, broke off from the Pine Island glacier. This is not the first time this has happened but should underscore our monitoring of the situation and the possibility of something far beyond the sea level rise projections that we are getting.

THE ARCTIC AND THE SNOWBALL EFFECT

When I was at Dickinson College studying geology in the early 70's, I was intrigued by the ice ages and their impact on sea levels. Back then, just four decades ago, it was inconceivable that the Arctic would be ice-free in the twenty-first century. The Arctic Ocean has been frozen for almost three million years, spanning more than two dozen ice ages, as well as the warmer eras in between, those "interglacials," like the period we live in now.

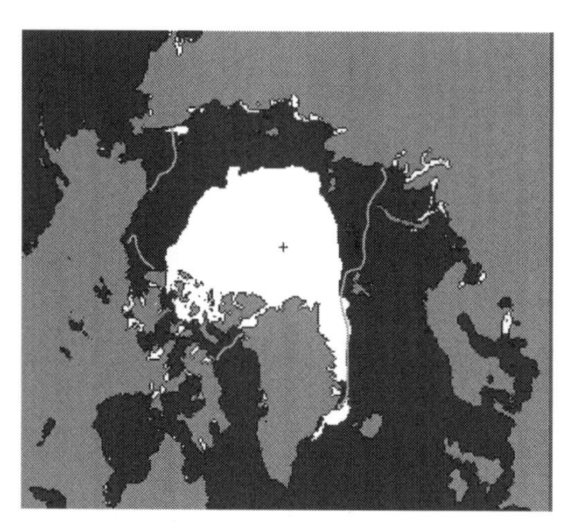

Figure 7-5. 2012 NASA / NOAA satellite image shows the decline in multi-year ice in the Arctic Ocean, surrounding the North Pole. The line circling the pole shows average annual ice extent over the past two decades. For orientation, North America is lower left, Greenland lower right, and Russia at the top of the image. (Graphic, courtesy of the National Snow and Ice Data Center, University of Colorado, Boulder.)

If I had suggested the idea of an ice-free Arctic to my professor, the late Dr. Henry Hanson, I have absolutely no doubt that he would have said something like, "John, that's a really interesting theory. But unless you can come up with any kind of plausible explanation as to how the polar ice cap could melt during the next century, you may want to consider another major, perhaps in the English department, writing science fiction." An ice-free Arctic was literally not even a topic of discussion among geologists.

For the last three decades, the Arctic sea ice has been melting quickly. Recent estimates for when the Arctic Ocean will become largely open water each September range from as early as this decade to as late as the 2040's.[39; 40] In fact, in September, 2012, Arctic researcher Dr. Peter Wadhams from Oxford University stated that, based on the unprecedented melting that summer, the Arctic could be ice free as soon as 2015 or 2016.[41] Fortunately in 2013 the Arctic sea ice rebounded considerably from the 2012 record low. Nonetheless, Figure 7-6 shows that such up and down variations of size are normal even with the strong downward trend.

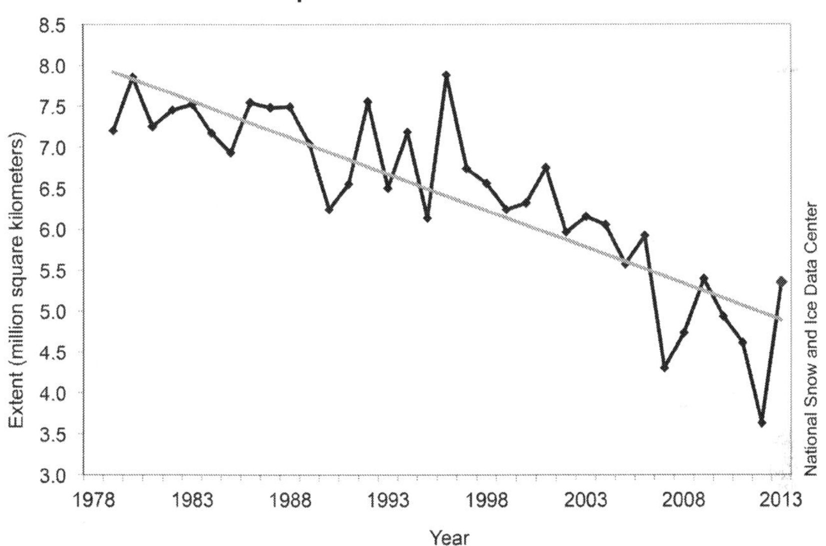

Figure 7-6. Plot of the annual changes in floating Arctic sea ice. 2012 had a record decrease in the polar ice cap, that rebounded sharply in 2013, similar to the recent decades. The 30-year trend is very clear. More images and information may be found at www.arctic.noaa.gov and www.nsidc.org.

Eskimos, or more accurately, the Inuit, say that Arctic weather in recent years is *uggianaqtuq* (OOG-e-a-nak-tuk), which means, "acting unexpectedly, in an unnatural way."[42] Their cultural memory and oral history are a good reinforcement to the modern scientific insights. (For more about our changing climate and influence of the high Arctic, I strongly recommend Dr. Heidi Cullen's book, *The Weather of the Future.* Heidi is a meteorologist and climatologist, previously on The Weather Channel and now an expert at Climate Central.)

The Inuit are expert at reading the ice, discerning a lot of information from nuances that only they notice. In 1985 I led a diving expedition under the polar ice cap, south of Canada's Ellesmere Island. Despite the renowned expertise of our lead guide, we lost a snowmobile through the ice, a bit of a disaster. The ice had thinned much earlier than usual. "Uggianaqtuq", they muttered. They were embarrassed, but also perplexed. That was 25 years ago, and apparently the patterns were just starting to change.

While the melting Arctic ice cap represents a profound and disruptive change, it has almost no direct effect on sea level rise because, as mentioned at the beginning of this chapter, the Arctic ice cap floats on water, not on land. Think of a large ice cube floating in a glass, with about ten percent rising above the liquid. As it melts, it does not change the level of the liquid at all; this is known as Archimedes Principle.

Nevertheless, the loss of the Arctic ice sheet will have a powerful effect. For example, an ice-free Arctic will bring forth profound changes in global weather patterns. The bright ice, frozen year-round, acts as a giant reflector, returning more than 70 percent of the sunlight heat energy back into space. If the ice is covered with snow, the reflectance can be as high as 90 percent. This reflection has been part of the planet's energy and heat balance for several million years.

When the bright snow and ice melts and is replaced by very dark seawater, it becomes a heat-absorbing surface, reflecting only about six percent of the sun's energy. The result is that millions of square miles of Earth's surface are absorbing 10 to 15 times as much heat energy as it had when covered with snow and ice. This speeds up the warming process already underway, which increases the amount of water vapor in the air.[43] Water vapor increases the heat in the atmosphere, which then increases the rate of ice melt. This is what scientists call a "positive feedback loop" or, in non-scientific terms, the "snowballing" effect. Ironically, in this case snowballing means disappearing snowballs, and more.

Our awareness of the changes we are living through is only just emerging. One of the most widely respected Arctic authorities is Dr. Mark Serreze, Director of the National Snow and Ice Data Center in

Boulder, Colorado. He recently stated that the "Arctic is in a death spiral."[44] With the utmost sincerity he said to me, "Just a decade ago, if you had asked me about global warming and the melting Arctic, I would have been skeptical, believing that it could still be within the realm of natural cycles. But now I am convinced that we are witnessing something that will likely result in an ice-free Arctic in my lifetime. It still is hard to accept. We need to continue to study it, but what I have witnessed in these last few years has convinced me that these are not natural cycles in the normal sense of what has occurred during the last hundreds of thousands of years."

Beyond the question of when the Arctic ice will fully melt is the question of when it might return to "normal," in other words, stay frozen year-round with something resembling the polar ice cap that we have taken for granted.

I put that question to Dr. Alexander "Sandy" MacDonald, Director of the National Oceanic and Atmospheric Administration's (NOAA) Earth Systems Research Laboratory in Boulder, Colorado, and a leading professional in the field of atmospheric research. He told me, "In the context of our current models and the historical record, it appears as if it will be at least 1,000 years before the Arctic is permanently frozen year-round again, perhaps much longer."

Other experts agree. In his excellent and very concise book, *The Long Thaw*, climate scientist Dr. David Archer goes even further when he states that we are unlikely to return to the present level of ice coverage for 100,000 years, due to the extremely long time that carbon dioxide persists in the atmosphere.[45] Yet, even Archer agrees that while time is running out, we still can avoid a worst-case scenario if we make the right decisions soon.

GLACIERS—FRESH WATER FORCES

Glaciers have immense power to grind and shape terrain. They directly add to sea level as they melt or calve into icebergs that displace seawater. Glaciers exist on every continent except Australia. About 72,000 glaciers are now monitored by digital images through the World Glacier Monitoring Service. Their 2008 survey concluded that, "the average annual melting rate of glaciers appears to have doubled in the last decade, with record losses posted in 2006 for a key network of reference sites."[46]

In Montana, the shrinking namesakes of Glacier National Park were predicted to be gone by 2030.[47] Then, in March 2009, the park's resident expert, Daniel Fagre, was quoted in *National Geographic News* saying the glaciers will "be gone ten years ahead of schedule," in 2020.[48] While portions of the world's highest glaciers, such as in the Himalayas, will persist for centuries, many others will disappear in the coming decades.

Watching the world's glaciers retreat is stunning, if not sad. Yet the melting of all the world's glaciers—from Alaska to the Andes, the Alps, and the Himalayas—will only raise sea level by about one or two feet (50 centimeters). Recall that it is the melting of the two vast Greenland and Antarctic ice sheets that will dramatically raise sea level, not the glaciers. While the glaciers' disappearance will just have a modest effect on sea level rise, it will have an enormous impact on vast areas surrounding them. Extreme melting will initially cause excessive runoff followed by disastrous flooding, erosion, and crop damage. When the glaciers are completely melted, there will be a dramatic drought, cutting off a critical fresh water supply that had existed for thousands of years.

More than 100 million people in South America and more than a billion people in Asia rely on glacial runoff for all or part of their fresh water supply.[49] As glaciers shrink, people will be forced to emigrate in search of water and arable land. This problem is likely to hit within mere decades.[50] While it is certainly true that glaciers fluctuate naturally, the disappearance of most glaciers has not occurred on our planet for many millions of years. We have zero experience with this environmental/agricultural event.

More than 99 percent of sea level rise will ultimately come from melting just the two ice sheets on Antarctica and Greenland. Unlike that ice on land, the melting of the floating polar ice cap and the glaciers does not raise sea level, but is a good indicator of warming temperature.

Now that you understand the significant components of sea level rise, it's time to take a look at what causes the melting.

Chapter Eight
CO_2 and You

Satellites from NASA and from other nations show that global average air temperatures are rising. While year-to-year changes in temperature and precipitation can be confusing, any single year is irrelevant because of the complexity of factors that could cause a confusing "blip." A good timespan to observe climate trends is five to 10 years. As decades meld into a century the long-term temperature pattern becomes even clearer. As mentioned earlier, the one exception to this is when tipping points come into action, which we will address shortly.

THE INVISIBLE HEAT TRAP

Today we are bombarded by information about carbon emissions. While the subject is new to many of us, the science of carbon dioxide's effects on atmospheric temperature goes back almost two centuries.

In 1826, an extraordinary French scientist, Joseph Fourier, demonstrated that carbon dioxide was transparent to light, but trapped heat in the atmosphere, a phenomenon we now know as the greenhouse effect. (Fourier is more commonly identified with a brilliant mathematical equation.)

Seventy years later, in 1896, a Swedish scientist named Svante

Arrhenius wondered what would happen to atmospheric temperature if the amount of carbon dioxide doubled. His interest may have been spurred by the fact that it was the era of Charles Dickens, when London and other cities were blackened from the smoke of burning vast amounts of dirty coal. It took him a year to do the calculation with pencil and paper.

The Arrhenius analysis predicted that if the level of carbon dioxide doubled, from 280 ppm to 560 ppm, average global temperatures would increase approximately nine degrees F (five degrees C). Amazingly, his calculation is roughly comparable to modern predictions using satellites and supercomputers. While this is in part testament to the extraordinary quality of his work, it also points to the fundamental physics of this issue. Though invisible, carbon dioxide's powerful heat-trapping effect is straightforward physics and chemistry.

Global CO_2, Temperature and Sea Level

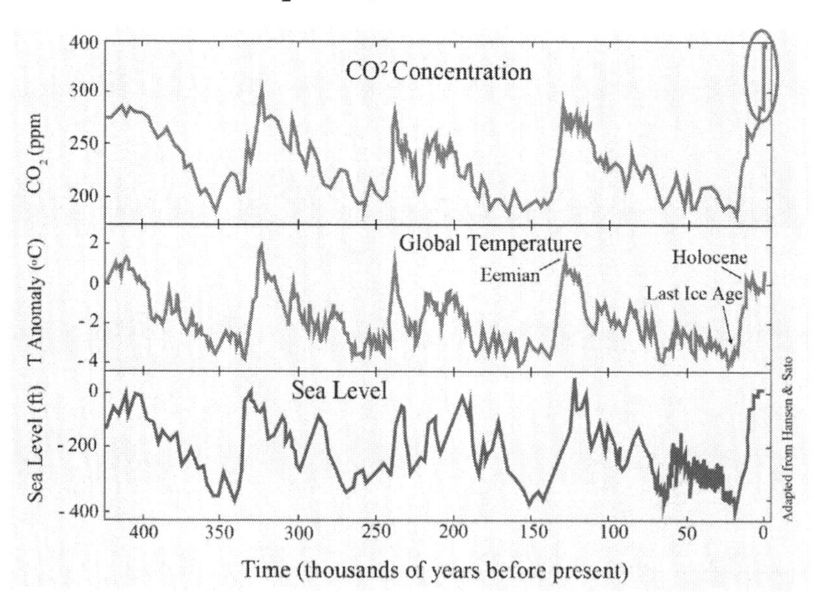

Figure 8-1. Sea level, temperature and carbon dioxide over the last 400,000 years shows the strong correlation of the three factors. Note that the carbon dioxide line (at the top) spikes up vertically at the end of the graph, reflecting the current level of almost 400 ppm and rising quickly. (Graphic by Hansen and Sato.)

Carbon dioxide is an extremely effective insulator. The amount of atmospheric carbon dioxide is now approaching 400 ppm and getting higher every year. Even at 400 ppm, that is only 0.4 percent of the atmosphere. Physicists calculate that if we had no carbon dioxide in our atmosphere, temperatures on Earth would be about 58 degrees F (32 degrees C) colder. [51]

Remarkably, we don't yet feel much of the extra heat resulting from the carbon dioxide blanket, because about 90 percent of the extra heat in the atmosphere is absorbed by the oceans. They act like a giant heat battery[52] due to the fact that water absorbs 800 times more heat than air.

Heat And CO_2: "Push Me, Pull You"

If you look carefully at the charts, you might notice that while carbon dioxide and temperature move up and down together, sometimes one leads, and sometimes the other does. Some climate skeptics have tried to use this inconsistency to claim that rising carbon dioxide does not raise temperature.

In reality, there is an amazing bidirectional relationship between the two. In other words, raising levels of carbon dioxide will raise the temperature, and rising temperature will raise the level of carbon dioxide in the atmosphere. While a surprising phenomenon, it can be explained rather simply.

We just reviewed how carbon dioxide traps heat, raising temperature. The reverse is even an easier principle of physics to demonstrate and understand. Oxygen and carbon dioxide in the atmosphere are mostly dissolved in a cool ocean. But when that ocean is heated it cannot hold as much dissolved gas. Think of opening two bottles of soda and warming one; the warmed bottle goes "flat" much more quickly because the heat forces it to release its carbon dioxide bubbles.

In our era, the increasing levels of carbon dioxide in the air are warming the atmosphere, which in turn will warm the ocean, which in turn will warm the atmosphere, and so on—another positive feedback loop that amplifies the situation.

In 2012 carbon dioxide levels were at more than 390 ppm and climbing at a rate of about three points per year. (For the current value, see the graphic in the right hand column of my blog's home page: www.johnenglander.net).

MAUNA LOA

The definitive study of carbon dioxide levels began in 1958, by Dr. Charles David Keeling. He established a laboratory on the Hawaiian Island mountaintop of Mauna Loa to minimize any possible effects from nearby civilization and to ensure that the carbon dioxide was "well mixed" within the atmosphere. Keeling invented a way to precisely measure carbon dioxide at an accuracy of better than one part per million. His rigorous, disciplined measurements are undisputed, even by those who question the causes and impacts of climate change (see Figure 8-2).

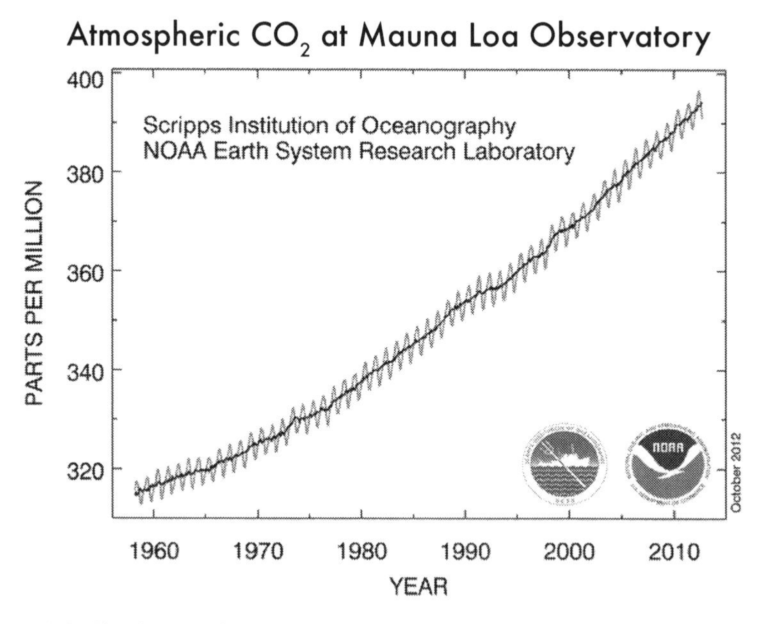

Figure 8-2. Charles David Keeling's record of carbon dioxide levels, taken between 1955 and 2004, has proven to be one of the most important trend records relating to climate. He found a way to measure carbon dioxide with incredible accuracy. These findings have since been verified by other laboratories in many other parts of the world and continue as a long term index of this critical greenhouse gas. (Graphic, courtesy of esrl.noaa.gov/gmd/obop/mlo/.)

The precise carbon dioxide measurements recorded over the last 60 years at Keeling's laboratory demonstrate that the carbon dioxide level is going up very steadily. They also show a very consistent sawtooth temperature pattern that correlates with the seasonality of vegetation cycles. The greater land mass in the northern hemisphere consumes more carbon than the southern hemisphere, accounting for the annual seasonal reduction.

As Figure 8-3 shows, when the Mauna Loa measurements are compared with records of carbon dioxide from preserved polar ice cores, the enormity of the recent change is clear and alarming. Given the hundreds of thousands of years that sea level, global average temperature, and carbon dioxide levels have moved in complete synchronization, the question is how quickly temperature will adjust to rising carbon dioxide, and then how quickly sea level will adjust to rising temperature and melting ice sheets.

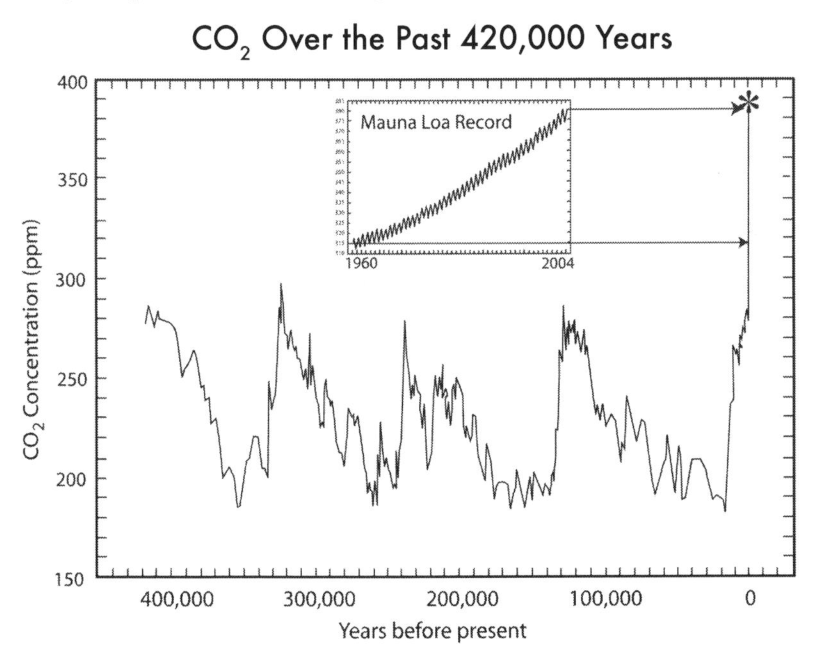

Figure 8-3. Putting Keeling's carbon dioxide measurements of the last half century on the historical record of 420,000 years shows the magnitude and suddenness of the change. Carbon dioxide is now approaching 400 ppm, a level that has not existed for 15 million years. (Graphic, courtesy of NOAA.)

As we have seen, sea level and climate have changed wildly over Earth's history, before there was any effect from humans. Over hundreds of millions of years, the level of carbon dioxide in the atmosphere ranged from hundreds to thousands of parts per million. Over the last few million years of the ice age cycles, carbon dioxide had a much more limited range, from 180-280 ppm, tracking the ice ages.

$C0_2$ Changes: Caused By Nature Or Man?

Many natural factors affect climate, including the annual cycle of the seasons, the amount of vegetation, volcanic activity, changes in the El Niño ocean current and solar cycles. The question of how much these influence warming compared to human activities that release carbon dioxide has been evaluated by thousands of scientists globally.

Figure 8-4 presents a composite of key graphs that show the influence of various factors on climate change. This is one of those images worth a thousand words. Look at each of the forces that affect temperature: solar cycles, greenhouse gases, volcanoes, and the El Niño cycles. The only one that significantly follows the trend of global temperature is greenhouse gases.

This and other analyses make it clear that while other factors can cause temporary short-term changes in the climate, they simply do not have sufficient force to change global average temperature for any sustained length of time.[54]

For example, volcanoes put vast quantities of carbon dioxide into the atmosphere, but show only a brief blip on the graph. It is true that when they erupt, massive quantities of water vapor, carbon dioxide, and other gases are released.

But all of those emissions dissipate quickly and are quite small compared to the human-generated emissions of the last century. The U.S. Geological Survey recently reported that, "On average, human activities put out in just three to five days the equivalent amount of carbon dioxide that volcanoes produce globally each year."[55]

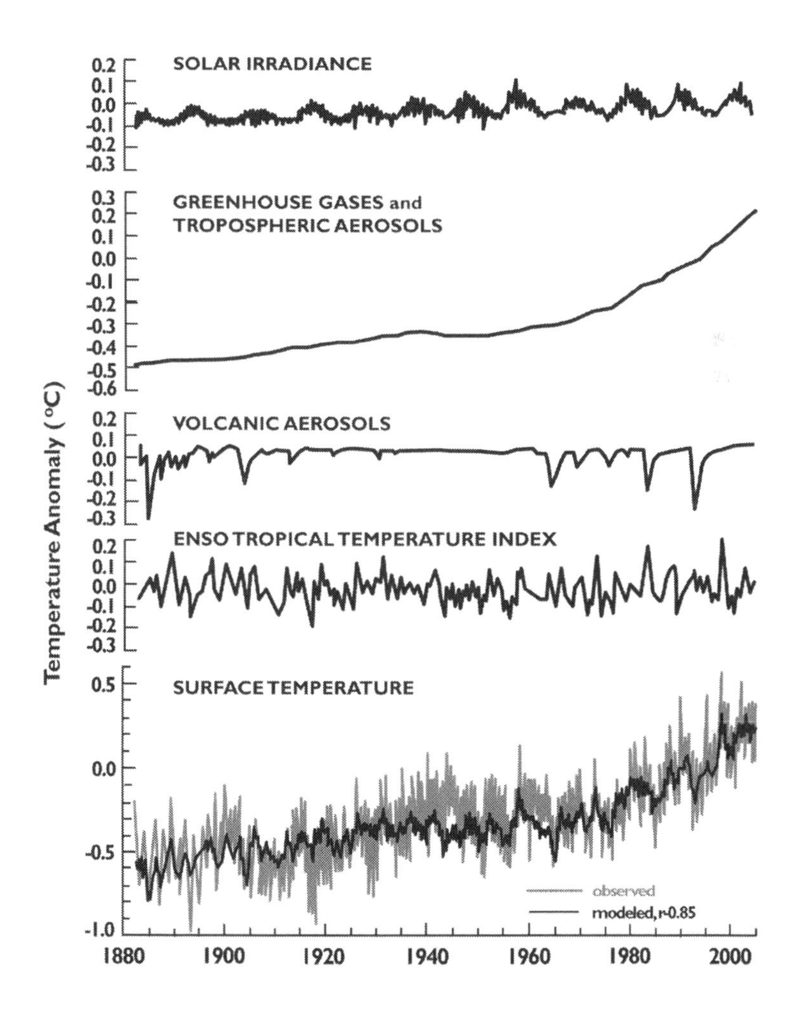

Figure 8-4. These five graphs provide a quick visual analysis of the components of rising temperature. Each chart covers the same time period, from 1880 to 2000. The warming is shown at the bottom. Above are four heat factors to which some people have attributed the warming: solar cycles, greenhouse gases, volcanoes, and El Niño ocean temperatures. Greenhouse gas is the only one that resembles the temperature rise graph. While not conclusive by itself, this one picture is worth a thousand words of explaining the century of data. (Graphic, courtesy of Lean and Rind, 2008.)

In addition, the cooling effect of volcano-released sulfur dioxide can even be stronger than the warming from carbon dioxide, though that varies over time. For instance, in 1991 the eruption of Mount Pinatubo in the Philippines was the largest volcanic event of the century. It caused a short-term global cooling of 0.9 degrees F (0.5 C).[56] Another example is the "year without summer" in 1816, caused by the massive volcanic eruption of Mount Tambora. Average global temperature dropped one degree F (0.5 C), causing crop failures and thousands of deaths from the cold.[57]

This cooling effect of sulfur dioxide has led some to believe that we could combat the atmospheric warming by putting sulfur dioxide into the atmosphere to counteract the effects of carbon dioxide. As we will see, this type of geo-engineering is simply not realistic.

But I digress. The takeaway message here is that temperature, carbon dioxide, and sea levels move in long-term synchronization. Temperature and carbon dioxide have a bidirectional sympathetic relationship, meaning the movement of one, up or down, will cause the other to do the same. Of the three, the primary variable that humans directly affect is the amount of carbon dioxide in the atmosphere.

Chapter Nine
Sources of Confusion

Donald Trump observed the record cold and extreme snowfalls in the winter of 2010 and publicly declared that Al Gore should lose the Nobel Prize for his film about global warming.[58] While the media ate it up and it was worth a chuckle, the comment unfortunately reinforced public misinformation. If Trump ever tries to add climatologist to his extensive resume, he should be on the receiving end of his most famous phrase, "You're fired."

As you now appreciate, more snowfall in one year does not mean the world is getting colder. In fact, a warming world will see a lot more moisture from a warmer ocean, causing significantly more precipitation overall. For decades to come that will include more snow in some areas, and in other areas lots of rain; in others, drought.

Misunderstandings about snowfall and many other changing weather patterns may be used to challenge the facts about climate change. Remember, it is important to look at five to 10 year trends rather than yearly weather data.

Weather Experts Versus Climate Experts

In "The Donald's" defense, even many television weather "experts" are confused about climate change. They are more properly referred to as "broadcast meteorologists," and a recent survey by George Mason University found that 25 percent of them say global warming is not happening, and 21 percent say they don't know if it is. Almost half said they needed some or a lot more information before forming a firm opinion about it.[59]

The problem goes well beyond the comedic parodies about some weather forecasters chosen for their looks or personalities. Traditionally, even certified weather broadcasters were not trained in climate issues. The highly respected *Columbia Journalism Review* ran its January 2010 cover story on this very issue with, "Hot Air, Why Don't TV Weathermen Believe in Climate Change?"[60]

Most people assume that weather experts have expertise in climate science because of the apparent similarity in the two fields of study. After all, climate is essentially long-term weather patterns, right?

Not true. Meteorology and climatology are vastly different. To appreciate the distinction, first consider other professions.

Think about accountants and economists. Both professionals deal with finances, yet their skill sets are entirely different; one is "macro" and the other "micro." The person who analyzes and prepares financial statements for a company would not know where to begin analyzing the national economy; the reverse is probably true as well.

Similarly, meteorologists study and predict the weather days and weeks from now. Climatologists look at climate patterns and norms over many years, centuries and millennia. The knowledge, methods, and tools used by each are totally different.

To be fair, there are a number of meteorologists who have gone on to broaden their scientific understanding and have developed real expertise in the field of climatology. In fact, some meteorologists have become leading researchers and experts in the field of climate change.

Thankfully, the principal professional meteorological organization, the American Meteorological Society, has recognized the need to better educate weather experts on climate issues and is now working to make this a high priority.

Forecasts Depend On Assumptions

Confusion also arises around predictions for the rest of this century. Mainly, this has to do with the variety of scenarios that are portrayed by computer models. To arrive at these scenarios, scientists must make assumptions about such variables as population levels, the amount of energy that will be produced, and whether that energy is generated by technologies that minimize or eliminate carbon dioxide emissions.

No one can possibly know all the answers to those questions today. Experts can only make intelligent estimates, and usually do so with a range of high and low numbers. By combining the various estimates for different factors, they typically run the models with several different scenarios, or sets of assumptions.

For example, Figure 9-1 shows the predicted temperature increase for this century, shown in degrees Celsius. (To convert one degree Fahrenheit multiply by 1.8; doubling is a rough estimate.) Each scenario has a more lightly shaded area, which fans out over time as the effects of the variables play out differently. While this type of graph has been used by skeptics to ridicule climate science, in fact, such diverse predictions prove that the science is sound.

On the other hand, projections that look out just a few decades can be even more precise. Thus the range of projections at 2050 is much smaller.

IPCC Report Misleading

The Intergovernmental Panel on Climate Change (IPCC) consists of several thousand of the top climate scientists and reviewers from over a hundred nations. Since 1988 it has issued a major report every five or six years about all aspects of climate change, including sea level rise. Generally, this encyclopedic document is the definitive reference authority in this field.

Unfortunately in the case of sea level rise, the IPCC has adopted an approach that many scientists, including this author, find to be badly misrepresentative. While I have the greatest respect for the IPCC overall, their way of presenting the data about future sea level falls short.

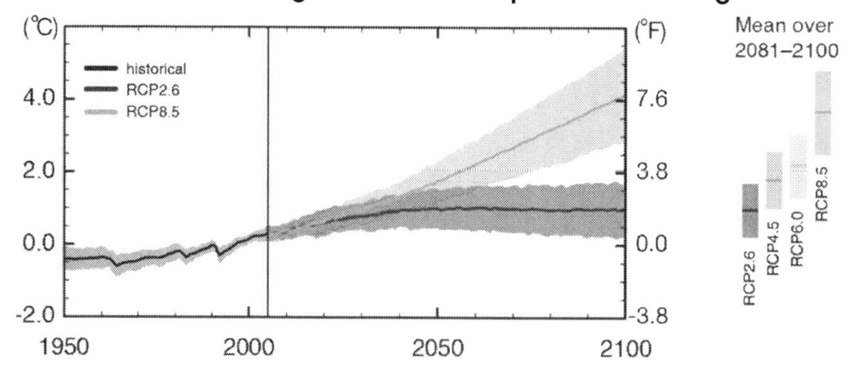

Figure 9-1. Projections about global average temperature increase are made according to various scenarios, based on variables such as population growth, energy use per person, and technology used to produce the energy. This graph from the IPCC shows the temperature increase that is projected with four different scenarios. (Source, Climate Change 2013: The Physical Science Basis. Working Group I contribution to The Fifth Assessment Report of the Intergovernmental Panel on Climate Change, *Figure SPM.10. Cambridge University Press.)*

On September 27, 2013 they released the first part of their Fifth Assessment Report.[61] Most journalists, elected officials, and even many scientists do not really understand what is behind the simple figures highlighted in "tables" by the IPCC. It is important to be aware of the IPCC process and why it may seriously underestimate sea level rise.

The 2013 report does raise the summary estimates for sea level this century to 10 – 32 inches (26-82 cm) as shown in the table in Figure 9-2, considerably higher than the 2007 figures of 7 – 17 inches (18-59 cm). Not surprisingly most readers take the mid-point of the ranges to be a good estimate. For the latest figures that would be 21 inches (54 cm) of sea level rise by the year 2100. While even that would be quite disruptive for our coastal society and economy, many of us believe that it is not only low, but misleading to the point of being irresponsible.

To get the true story you have to understand the protocols of the IPCC, read the fine print of the report, and know how to read "between the lines." Briefly the problems are:

IPCC rules dictate rigorous requirements for inclusion, as they should. Projecting how high sea level will rise by the year 2100 requires determining a specific number of centimeters for that year and getting a jury of peers to agree to it through the process of independent publication in a scientific journal. Generally scientists only put their names on projections for which they have very high confidence. For example, in the case of the *overall conclusions* of the latest IPCC report, they assess their confidence at 95 percent. That cautious approach creates an extremely "high bar" or filter for what can be included. As a result a lot of very good and alarming information is almost ignored.

Projected Sea Level Rise per IPCC 2013

Note Author's Cautions About This Data In Accompanying Text

Years	2046-2065			2081-2100		
	Likely Range - Inches (cm)			Likely Range - Inches (cm)		
Scenario	low		high	low		high
RCP2.6	6.7" (17cm)	to	13"(32cm)	10"(26cm)	to	22"(55cm)
RCP4.5	7.5" (19cm)	to	13"(33cm)	12"(32cm)	to	25"(63cm)
RCP6.0	7.1" (18cm)	to	13"(32cm)	13"(33cm)	to	25"(63cm)
RCP8.5	8.7" (22cm)	to	15"(38cm)	18"(45cm)	to	32"(82cm)

From AR5 WG1 Table SPM2

Figure 9-2. The recently released IPCC 1013 report shows projected sea level at mid century and end of century according to four different scenarios or sets of assumption. Unfortunately this simplistic presentation misses all the potential contributions to sea level rise that do not meet the very demanding criteria for high probability and specific quantity. In other words there are some "tipping points" that are not considered in these figures, creating a misleading impression that 32 inches (82cm) is the worst case projection.

The current table of sea level projections for this century largely discounts three phenomena that are believed to be nearing "tipping points" – those sudden departures from recent trends:

1. The melt rate in Greenland has been accelerating dramatically in the last decade, but varies from year to year. As a result it is still not possible to extrapolate precisely how bad it will be 87 years from now. To be responsible the IPCC approach uses a figure that assumes nothing dramatic changes in Greenland from the recent

pattern. As previously explained, Greenland holds the potential to raise sea level more than 20 feet.

2. Antarctica is showing ominous signs of melting just as Dr. John Mercer predicted in 1978, when he described a possible catastrophic sea level rise during the middle of this century. Recent data confirms the possibility of as much as 10 feet of sea level rise in a rather short number of years, but it does not meet the test of numerical specificity and ultra-high probability, so it too is essentially ignored by the summary IPCC projections.

3. Methane can be a game changer, a truly huge accelerator of warming and ice melt, as we will cover shortly. But again, the inability to quantify how bad it will be and whether it will surely happen before the end of this century is the reason the IPCC omits it from the sea level projections, leaving the subject "for further study."

In short, the IPCC summary sea level rise projection for this century is based primarily on the measurable melting of glaciers, the current melt rate of the Greenland Ice Sheet, and the thermal expansion of seawater. That misses about 95% of the potential problem. In the actual text, the 2013 IPCC report acknowledges many of these factors but does so in the cautious language of committees and consensus seekers.

To quote from their report:

> *Based on current understanding, only the collapse of marine-based sectors of the Antarctic ice sheet, if initiated, could cause global mean sea level to rise substantially above the likely range during the 21st century. However, there is medium confidence that this additional contribution would not exceed several tenths of a meter of sea level rise during the 21st century.*
>
> *The basis for higher projections of global mean sea level rise in the 21st century has been considered and it has been concluded that there is currently insufficient evidence to evaluate the probability of specific levels above the assessed likely range. Many semi-empirical model projections of global mean sea level rise are higher than process-based model projections (up to about twice as large),*

but there is no consensus in the scientific community about their reliability and there is thus low confidence in their projections.

[From IPCC AR5 WG1 SPM-18]

Most readers and journalists miss these nuances. From the previous 2007 report most readers noted that the highlighted figures showed an average of only a foot of sea level rise this century, not realizing it entirely omitted all the big uncertainties or "wild cards" cited above.[62; 63; 64]

The 2013 report raises that mid-point estimate to two feet, double the earlier one, but is still conservative. Articles are describing a worst-case scenario of "almost three feet" with some adding, "possibly more." In fact three feet is not the worst case at all.

For example, one independent team of highly regarded experts, led by Dr. Stefan Rahmstorf, estimates a mean sea level rise this century of four feet (1.2 m) and a high range of six feet (1.9 m). Even that amount does not take into account the big tipping points or wild cards, particularly the destabilization of West Antarctica.

I do appreciate the IPCC's dilemma. They do not want to be alarmist. Their credibility is at stake. Yet there is so much at risk with rising sea level, their cautious approach to presenting the forecasts likely misleads the world to significantly understate what could happen this century.

LAG TIMES — SWIMMING POOL EFFECT

Because water is 800 times denser than air, it takes a lot more heat to raise the temperature of water than it does to warm the same amount of air. An outdoor, unheated swimming pool illustrates this. While air temperatures can change significantly from day to day, water in a pool will take several days to adjust its temperature either up or down, depending upon the temperature differential and the depth of the pool. The same applies to our very big pool—the ocean, which is up to seven miles deep.

To explain what I call the "swimming pool effect" in a tad more detail, the oceans have different temperature layers, known as thermoclines, which are very distinct. You may have felt them swimming or diving. These boundaries slow the mixing of temperatures. It can take decades even for the surface layer—the top 600-700 feet (200 meters)—to fully adjust to a single degree of warmer air temperatures. Pioneering oceanographer Dr. Wallace Broecker estimates that it could take as long as 1,000 years for the entire ocean to equalize to a new average global air temperature.[65]

This is one of the reasons that sea level is projected to rise for at least 1,000 years after the atmospheric temperature stabilizes again. The excess heat in the atmosphere is still warming the ocean. Even when it stops that excess ocean heat will continue to keep the atmosphere warm and continue to melt the ice until it reaches a new point of equilibrium. As a point of reference, the historic global average temperature at sea level has been 59 degrees F (15 degrees C) and is already one degree higher.[66]

Those who simply dismiss the issue of climate change as something that can be solved by the next generation do not take into account this huge lag time. What we do today will have cumulative, amplified effects on our children, grandchildren, and beyond.

It strikes me that there are similarities between this problem and the national debt. The longer we delay dealing with each crisis, the worse the pain of the solution, which inevitably must come. A key difference is that climate has a much slower response than financial systems. Unfortunately, that slowness makes it even more tempting to delay dealing with the rising temperature, but also means the remediation will be much slower, too.

After one lecture, someone asked me why we could not just "cool the ocean." The fact that this question was met with some laughter indicates that most people appreciate the herculean task of removing that much heat energy. However, in the spirit that any idea should be considered, I responded, saying, in essence, two problems: First, if you could remove the heat, as we do with a refrigerator, where are you going to put the heat energy? Our refrigerators typically dump the heat into the kitchen; air-conditioning usually then moves that heat outdoors. Second, just as it takes energy to operate a refrigerator or air conditioner, it would take mind-boggling amounts of energy to remove significant heat from the ocean. And of course, the way we got into this situation is that our energy generation produced the greenhouse gases that raised the temperature. Cooling the ocean by any energy-driven device is not an option with current technology.

Confusing Temperature References

We are hearing more predictions of temperature increases over the course of this century that can be confusing. International leaders announced a goal to keep planetary warming to no more than 3.6 degrees F (2 degrees C). Following the Copenhagen Accord in December, 2009, that target was affirmed by the "G8" and the "G20," the two consortia of major nations.

That temperature goal assumed that the groups meet all the emissions targets for greenhouse gases by about 2015, but such emissions targets have not even been agreed upon. Countries or regions like Europe that did set targets mostly failed to achieve them.

To put the reality of that goal in perspective, a recent estimate from Massachusetts Institute of Technology (MIT) anticipates between seven and 11 degrees F (four to six degrees C) of warming this century if we stay on our current path of carbon emissions.[67] If that nine-degree F mid-point does not sound too bad, keep in mind that the Global Average Temperature (GAT) is now only about nine degrees F (five degrees C) warmer than at the peak of the last ice age. Such temperature change would have devastating effects on all biologic systems, causing severe animal die-offs, droughts, agricultural devastation, and immense loss of human life.

Also, GAT includes combined temperatures of ocean and land, but the oceans do not change temperature nearly as much as land does. Land temperatures swing about 40 percent hotter and colder than the GAT. Thus, a nine degree warmer GAT temperature could mean 12.6 degrees F (7 C) higher temperature on land.

Finally, it's important to understand that temperature increases are much stronger at higher latitudes. Near the poles, the temperature increase is almost twice as large as the GAT. This warming will in turn melt more ice and raise sea level, accelerating the positive feedback.

Antarctica's Confusing Signals

A warming ocean means more precipitation. In Antarctica that means more snow. Up until about 2006 more snow was accumulating than was removed by melting, the collapse of ice shelves, or glaciers entering the sea.[68] The fact that East Antarctica was not melting seemed to contradict the assertions that the planet is warming. In fact, more snowfall on East Antarctica was exactly what a warmer ocean and more moist marine air were expected to create.

Regardless, the situation in Antarctica has now changed. In spite of increased snowfall, overall melting has outpaced snow accumulation, resulting in a net loss of ice. This accelerating melting is in accord with the recent climate models as well as Mercer's forecast almost a half century ago.[69]

Regional And Local Variations

Continents and islands move up or down, usually less than an inch per year. If the land goes up, sea level goes down relative to that coastline. The relative rising or sinking of a continent explains what might seem like inconsistencies in sea level rise in different locations.

Much of North America and Europe is still rebounding upward as a result of the reduced weight as the ice sheets disappeared. Alaska has so much uplift that sea level is actually going down relative to the shoreline.[70]

In other areas around the world, land is sinking. Land sinks when it is compressed, which can happen naturally from the compaction of sediments, or as a result of water or oil being pumped out of the ground.

Each of the above factors may only account for inches, or fractions of an inch of subsidence each year, but even such small amounts can create significant change in a particular location, when compounded by sea level rise.

Ocean currents can also affect regional differences in sea level. A good parallel for this is what happens with a swiftly moving stream. If you looked closely at such a stream, you might notice that the water level is slightly higher on the outer shore of a riverbend and slightly lower on the inner shore. Ocean currents have a similar effect on coastlines, though the effect is complicated by Earth's gravitational pull.

Wind produces similar results in that it can actually pile up water as it pushes it towards land. While the change in relative ocean height is only slight, and generally temporary, it can confuse short-term measurements of sea level.

Another factor is the gravitational pull of land masses. You may remember from school that objects attract in proportion to their mass; the greater the density, the greater the pull. Our planet is obviously not a homogeneous mass, but is quite "lumpy." So, places where density is greater will therefore have a greater pull on the ocean.

As a result the oceans are not smooth or evenly distributed. Assume for a moment that we eliminated the wind, the waves, and the tidal bulges. The ocean would still have hills and depressions, corresponding to the gravitational attraction of nearby mass. This is almost impossible to observe across the vast dimensions of the sea, but has been confirmed in the last few decades by precise satellite measurements.

This astounding concept was described in 2010 in the popular German magazine, *Der Spiegel.* Dr. Detlef Stammer explained that the height of different oceans can vary by hundreds of feet as a result of the uneven gravitational force in different parts of the planet.[71]

A fascinating result of this mass attraction phenomenon is that as the huge ice masses on top of Greenland and Antarctica melt, they will redistribute their mass, reducing their "pull" in those locations. Thus as Greenland melts in the north, it will disproportionally increase ocean heights in the southern hemisphere. The melting of Antarctica will do the opposite.

A 2009 study calculated that as the Western Antarctic collapses,[37] sea level on the U.S. coastline would rise 25 percent more than the global average. The same study showed that the redistribution of mass would also affect the earth's rotation, causing an increase in sea level along the North American continent and in the Indian Ocean. Fascinating, but frankly rather nuanced for this perspective.

SELECTIVE EXTRACTS FROM THE SEA LEVEL RECORD

As has been discussed several times now, sea level rise cannot be understood on a year-to-year basis. We must look at long-term trends. The longer the timeline, the clearer the trend.

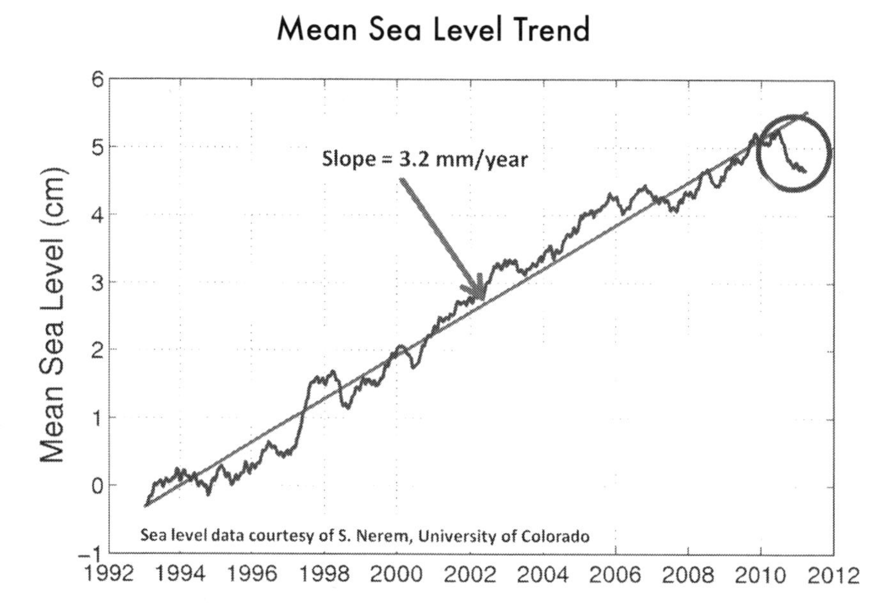

Figure 9-3. Sea Level will vary from year to year depending on the amount of melting snow and ice, as well as the amount and location of rainfall. While it can be unusually high or low for a year or two, the trendline becomes clear when looking at timespans of a decade ore more. (Graph, courtesy of S. Nerem, University of Colorado.)

This simple fact is often ignored in the mainstream media. For example, in 2011 many reporters seized on an article by NASA's Jet Propulsion Laboratory (JPL), which showed that for several recent years sea level was declining.[72]

Some reporters used this information to question the issue of sea level rise and climate change in general. What they failed to mention was JPL's description of the cause of the short-term drop. In the same report, scientists explained that it was caused by increased ocean warmth, which had led to more rain in certain areas. In turn, that rain caused an abnormally high amount of water to be stored in soils, lakes, and rivers. They noted that such a decline in sea level was temporary, expected, and would be limited by soil saturation.

Figure 9-3 is the graph that went with the story. While the decline is significant, there is no reason to believe that it is fundamentally different from the many other bumps over the last century. Nonetheless, these types of stories can be a major source of confusion.

Tide gauges show sea level relative to the location where the gauge is mounted. In some locations tide gauge data will show that sea level is dropping. This can happen if the land is rising (uplifting) at a faster rate than sea level, as is happening in Alaska, as one example. The uplift can be caused by a slow response to the removal of ice sheets, or can be a result of volcanic activity that is causing a bulge in the earth.

Disinformation Campaigns

There are some who have a stake in sea level rise not being a problem. Those who work for the fossil fuel industry could be an example. Perhaps they are not even aware that they have a bias to believe that burning fossil fuel is not damaging to our future climate, agriculture, weather patterns, and the coastline. This tendency was noticed nearly a century ago by Upton Sinclair, when he famously remarked, "It is difficult to get a man to understand something when his salary depends upon his not understanding it."

This is true at the corporate as well as the personal level. Fortunately, today there are some enlightened companies and senior executives who understand the devastating power of climate change. Just for example, the following companies have taken positions recognizing the threat of climate change and its association with the level of greenhouse gases: Alcoa, Boston Scientific, Chrysler, Coca-Cola, Dow Chemical, Duke Energy, DuPont, Exelon, Ford Motor, General Electric, Honeywell, Johnson & Johnson, Pepsi, PG&E, Rio Tinto, Shell Oil, Siemens, and Weyerhaeuser. There are many more. Awareness is changing quickly.

You may have noted that several companies are even involved in the energy business. I strongly believe that we should not make villains out of the companies that help meet our vital energy needs. They are providing a valuable service that we depend upon. Our power needs are not going to reduce magically, and the replacement technologies will take a lot of time to develop and implement, as I will cover.

Where I draw the line is when companies actively, even aggressively, create disinformation about carbon dioxide emissions to confuse public awareness or policy.

Cognitive Dissonance

One further reason for confusion about sea level rise has to do with the brain's ability to process bad news.

Until very recently, mankind has assumed the coastlines to be essentially unchanging, since they have changed little for the last 6,000-8,000 years. Our cultural heritage and our daily lives operate within a context of stable sea level. The boundaries of the land masses appeared to be as rock solid as anything we know, and had been so for all of our ancestors.

The magnitude of devastation related to sea level rise is so massive that we react with what psychologists term cognitive dissonance: when confronted with difficult and conflicting thoughts, we are uncomfortable and find some way to reconcile them.

We tell ourselves such things as: "It's just a cycle; cycles reverse themselves; it's just a matter of time; it will happen so far in the future, I don't need to worry about it; we can solve anything; the next generation will figure it out and fix the problem; leave it to them; we will just have to build seawalls."

Our discussion so far makes it clear that the first couple of rationalizations do not hold up. We will soon look at the fallacy of the other ideas.

CONSPIRACY THEORISTS/INCONCLUSIVE SCIENCE

Finally, there are those few individuals, including some politicians and media commentators, who want you to believe that scientists are in some vast conspiracy to fabricate data about global warming, as if there was some dark oath to collectively mislead the public as a scare tactic to shake down more funding for research.

Numerous surveys of climatologists have strongly refuted that assertion, including a 2007 survey of more than 10,000 geoscientists. Among those having an actual degree in climate science, 97 percent agreed that human-caused greenhouse gases are substantially affecting climate.[73]

One story, claiming that 31,000 scientists disagree with human-caused climate change, has been shown to be one of many internet myths. A thorough debunking of this fiction can be found at the excellent web site: www.skepticalscience.com.[74]

Anyone who believes that it is possible to get the majority of climatologists to intentionally falsify data in concert has not spent much time around scientists. By nature, scientists are challenging, independent thinkers. Such skeptics certainly have never tried to fool tens of thousands of the brightest graduate students on different campuses all over the world, who are working to advance the science and their own careers in an incredibly competitive environment.

While some may be motivated by a need to get a grant, this would be true for any scientific endeavor they would embark upon. In many cases, particularly in countries other than the U.S., the majority of scientists studying climate change issues have secure funding and simply no motivation to concoct nonsense to fund their research. The overwhelming majority of scientists work with integrity, and there is no reason to question the findings simply because they are disturbing. I personally know that many of the most passionate concerns are from extremely senior, even retired scientists, whose overriding focus is their children's and grandchildren's futures.

As is the case in all rigorous scientific arenas, the field of climate science ultimately polices itself through expert panels, via vigorous debate, and through peer-reviewed professional publications.

Perhaps the best confirmation of the reality of "global warming" came unexpectedly on July 28th, 2012, on the OpEd page of the *New York Times*. Dr. Richard A. Muller, a long-time champion of skeptics, wrote a game-changing column titled, "The Conversion of a Climate Change Skeptic."[75] The first two paragraphs are unambiguous:

> *Call me a converted skeptic. Three years ago I identified problems in previous climate studies that, in my mind, threw doubt on the very existence of global warming. Last year, following an intensive research effort involving a dozen scientists, I concluded that global warming was real and that the prior estimates of the rate of warming were correct. I'm now going a step further: Humans are almost entirely the cause.*
>
> *My total turnaround, in such a short time, is the result of careful and objective analysis by the Berkeley Earth Surface Temperature project, which I founded with my daughter Elizabeth. Our results show that the average temperature of the earth's land has risen by 2.5 degrees Fahrenheit over the past 250 years, including an increase of 1.5 degrees over the most recent 50 years. Moreover, it appears likely that essentially all of this increase results from the human emission of greenhouse gases.*

The fact that Dr. Muller is an esteemed scientist, who was openly skeptical about human-caused climate change, and that the infamous Koch brothers largely supported his work, gives it all the more credence.

Chapter Ten
Why Projections Underestimate

As you now know, information presented about climate change and sea level rise is often confused, if not downright misinformed. But even when this is not the case, it is often understated. Let's examine why.

SCIENTIFIC CAUTION

Generally speaking, a scientific approach is excellent at testing and debating new ideas. The peer review process ensures a high level of rigor in scientific study and analysis. Over time, the truth wins out, and occasionally significant new understandings emerge.

However, scientists by nature may be more conservative than one would assume. Their cautiousness often goes beyond a simple wish not to be alarmist, towards a bias to resist new ideas, or at least reluctance to stick one's neck out. This phenomenon is given little attention, though climate scientist Dr. James Hansen published a paper in 2007 titled, "Scientific Reticence and Sea Level Rise," in which he stated his growing concern about the caution of his profession:

> *Scientific reticence may be a consequence of the scientific method. Success in science depends on objective skepticism. Caution, if*

not reticence, has its merits. However, in a case such as ice sheet instability and sea level rise, there is a danger in excessive caution. We may rue reticence, if it serves to lock in future disasters.

I believe there is a pressure on scientists to be conservative. Papers are accepted for publication more readily if they do not push too far and are larded with caveats. Caveats are essential to science, being born in skepticism, which is essential to the process of investigation and verification. But there is a question of degree. A tendency for 'gradualism' as new evidence comes to light may be ill-suited for communication, when an issue with short time fuse is concerned.[76]

I have met with Hansen and can attest that he is a very modest, even humble person, perhaps not what one would expect from one of the world's leading scientists and the head of NASA's Goddard Institute of Space Studies for decades until his voluntary retirement in 2013. When this expert speaks, it is worth listening.

Hansen has a few critics, as do most pioneers. History makes clear that disruptive concepts inevitably face great opposition. Galileo, Copernicus, Darwin, Einstein, are well known examples of those who encountered resistance and ridicule along the path towards acceptance of revolutionary ideas.

As an example, many scientists believed that mass extinctions were caused by earth-based events such as volcanoes. Then, a radical paper in 1980 by the father-and-son Alvarez team proposed that the mass extinction nearly 66 million years ago that wiped out the dinosaurs and much of life on Earth was caused by a large asteroid.[77]

Over the course of a few years, a great portion of the scientific community came to accept this theory. Excavations all over the world showed the "signature" layer of iridium and other material consistent with a foreign body impact. When a huge crater was found in 1991 in the Mexican Yucatan its characteristics largely confirmed the evidence that had been building. Nonetheless, some scientists continued to deny the new concept.

The deniers said it was a coincidence, and held onto their theories of volcano-caused extinction. Some continue to voice disbelief to this day. In his fascinating book about extinctions and climate change, *Under a Green Sky*, leading paleontologist Dr. Peter Ward bluntly describes the battle among scientists to accept a new cause for extinctions:

> *But never count out the foes who just cannot afford to lose—massive reputations, massive egos were at stake... an increasing number of paleontological studies showing data consistent with a sudden extinction were opposed by the ever-decreasing doubters who were increasingly composed of cranks, the slow, and conservative, and those seeking attention by screaming in loud, if knowingly false, protest.*[77]

To be fair, a poorly executed study or a flawed theory can permanently tarnish one's professional scientific reputation and beckons caution. This conservative tendency may help to explain the observation of a 2010 study by Harvard economist Martin Weitzman. Using sophisticated statistical and economic analysis, he showed the bias to underestimate the projected damages from climate change. The paper's title suggests his conclusion, "GHG (Greenhouse Gas) Targets as Insurance Against Catastrophic Climate Damages."[78]

To summarize a few of his findings and observations:

- The damage that would be caused by 11 degrees F (six degrees C) of warming is considerably worse than what traditional cost-benefit analyses calculate.

- While the IPCC in 2007 said that the probability for climate catastrophe was effectively less than three percent, surveys of leading scientists indicate anywhere from a 9-50 percent chance of disaster on our current path of emissions.

- Six degrees (Celsius) of extra warming seems to be about the upper limit of what the human mind can envision for how the state of the planet might change. It serves as a routine upper bound in attempts to communicate what the most severe global warming might signify.

Scientific tendency towards caution is particularly pronounced when it comes to forecasts around tipping points. As described previously, a "tipping point" occurs when patterns change abruptly, dramatically, and sometimes permanently.

Tipping Points

A longer scientific view usually provides more accurate scientific data, but when approaching a tipping point, longer views can be deceiving. For example, Figure 10-1 from the National Snow and Ice Data Center[43] illustrates this with a dotted line that shows more dramatic change over shorter periods of time.

The figure shows the declining volume of floating sea ice in the Arctic. The decline in ice coverage is speeding up. As the two trend lines show, if we are truly at a tipping point, the shorter timeframe gives a better idea of the current direction.

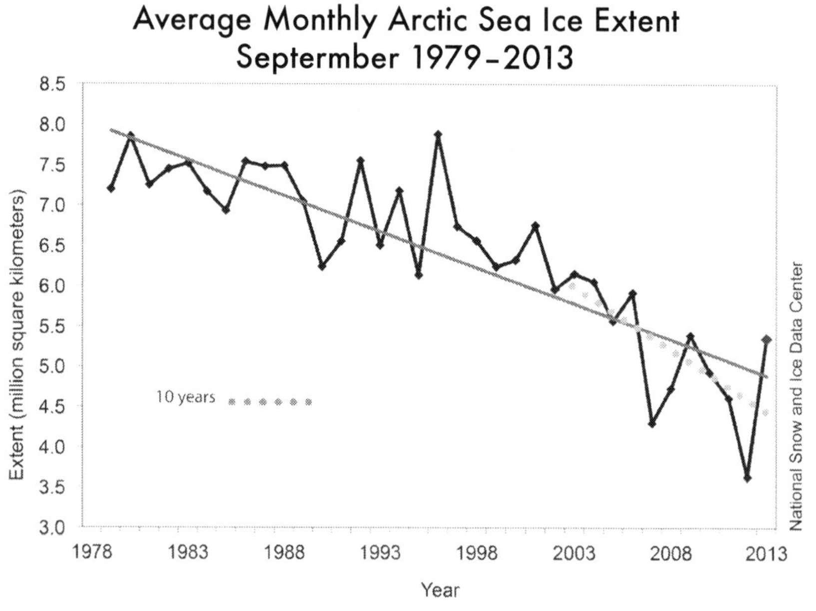

Figure 10-1. The long straight line shows the 30-year trend of the Arctic ice cap melting, smoothing out the variations of individual years. By also plotting a 10-year trend (the dotted line), we can see that the rate of melting is increasing. This illustrates a tipping point, a situation where the longer time period may give a less accurate picture of quickly changing trends. (National Snow and Ice Data Center, University of Colorado, Boulder.)

Once a tipping point is reached, it can be hard to know what pattern the new trend will follow. Since the Arctic has been frozen for millions of years scientists have not had experience examining precisely how catastrophic melting occurs. We have no choice but to rely on models and then see how well the actual observations confirm the forecast.

Figure 10-2 illustrates clearly that so far, the Arctic is melting faster than the models forecast, and that predictions for the coming years may be overly conservative. In Greenland, one of the two major contributors to sea level rise, ice sheets are melting faster than at any time in the last 100,000 years. Many believe it is also at a tipping point.

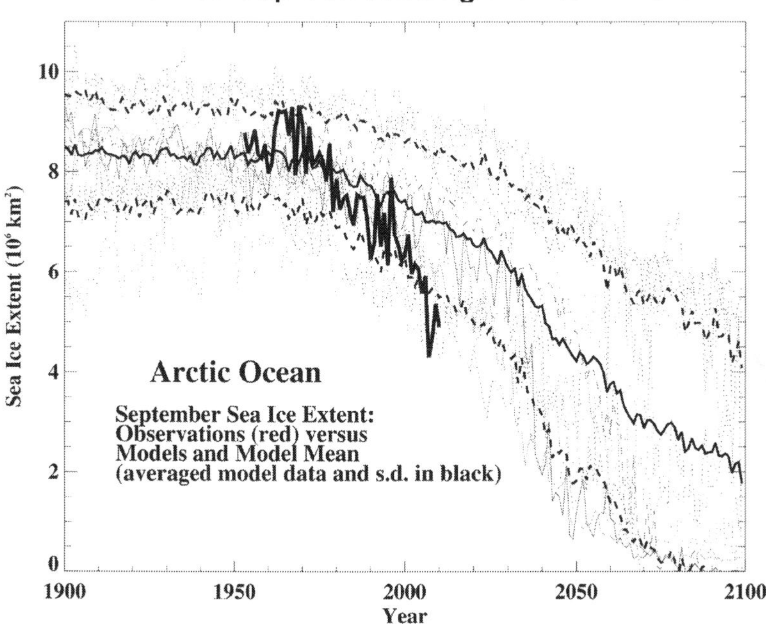

Figure 10-2. Two hundred years of models and projections for the polar ice cap surrounding the North Pole are shown by the several squiggly lines. The darker, tight, jagged line spanning from 1950 through the present shows the actual decline in the Arctic ice sheet. This shows that the Arctic is melting much faster than any of the models have predicteed. (Image, courtesy of the National Snow and Ice Data Center, University of Colorado, Boulder.)

Certainly, no credible model shows the Greenland melt rate slowing down this century. As referenced in Chapter Nine, the benchmark IPCC report uses several different scenarios with varying assumptions about population levels, energy consumption, and the use of different types of energy to predict what might happen to climate and sea level over the next century.

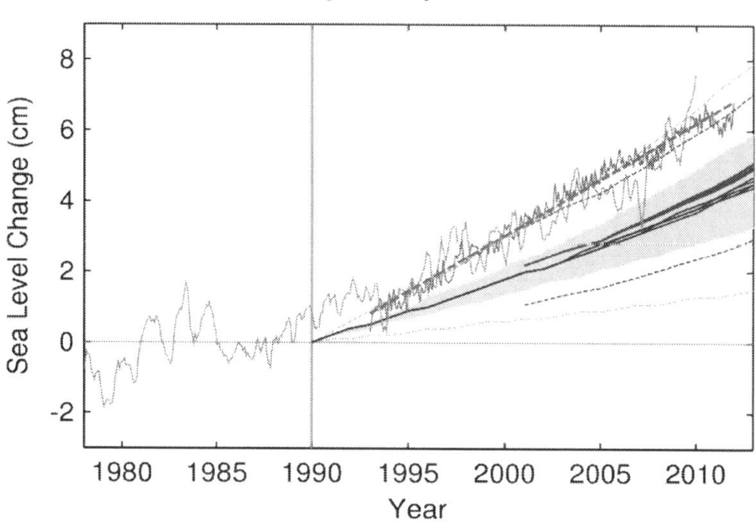

Figure 10-3. How good are SLR projections? The IPCC (Intergovernmental Panel on Climate Change) published SLR projections in 1990 and in 2002, shown above as teh fairly straight upper and lower boundary lines starting in 1990 and 2000. The squiggly line starting on the far left is actual sea level measured by tide gauges. The black line running through that from the mid 90's is satellite altimeter data. Note that the highest projections barely include the actual sea level. What this shows is that the IPCC projections for SLR are consistently low, a rather good indicator that future projections will continue the same pattern. (From Rahmstorm, 2012)

Figure 10-3 shows that sea level rise (the squiggly line) at the top is actually above the ranges of projections (the shaded areas and straight lines) for the past three decades. There is no way this should have happened, particularly just a few years into the future. Multiple prediction scenarios are intended to cover the entire range of all possibilities. This is clear evidence that most sea level projections are not aggressive enough. If anything, they underestimate the range of scenarios that may unfold.

Dr. Robert "Bob" Corell was first tasked to look at climate change by President Ronald Reagan in 1987, and has served as a senior climate advisor to every administration since, both Democrat and Republican.

Bob remains very engaged on the issue, recently as Chairman of the International Arctic Science Committee and currently as the lead author of the U.S. National Climate Assessment. Reflecting on many decades of studies about climate change, he observes, "The historic geologic record shows that we are under-projecting the future, not over-projecting, as the skeptics suggest."

Research by MIT scientists in 2009 showed that without stringent reductions of greenhouse gas emissions, the impact of climate change this century may be far more significant than some climate assessments have indicated. Their Joint Program on the Science and Policy of Global Change found significantly increased odds that global temperature increases would reach nine degrees F (four to five C) by the end of the century, much higher than similar studies completed just six years ago.[67]

20,000 Times Faster Than Nature

One of the challenges to precisely modeling ice melt and sea level rise this century is the unprecedented rate of warming.

The last truly abrupt changes in the earth's climate occurred more than 50 million years ago. During that period, carbon dioxide increased by about 100 ppm over a million years. The global temperature spiked by about nine degrees F (five degrees C) over 10,000 years.[79] While that may sound slow, in geologic time it is considered quick and drastic.

At our current rate of carbon emissions, we will increase carbon dioxide levels by that same 100 ppm in just 30 to 40 years. In other words, we are increasing carbon dioxide levels roughly 20,000 times faster than at any time in the last 540 million years.[80] Temperatures, which can lag behind the rise of carbon dioxide, are now rising about 55 times faster than they did even during the most recent cycle of glacial melting.

The rate and magnitude of atmospheric change are the key factors missing from the public discussion of climate change. As the "doubters" suggest, Earth's climate has changed before, including rapidly rising carbon dioxide levels. But the last time such an event happened, it took about a million years, and resulted in 75 percent of the species going extinct. The fact that carbon dioxide levels are increasing thousands of times faster than those past natural climate crashes should cause us to ponder this a little more thoroughly, rather than just to dismiss it as "having occurred in the past." The unprecedented rate makes it impossible to precisely predict the effects. As the sailors used to say, *we are entering uncharted waters.*

METHANE: THE BIG UNKNOWN

Of the 30 or so greenhouse gases, none is more ominous than methane. In its pure form it is more than 200 times more effective at trapping heat than carbon dioxide. Over a period of years it breaks down into carbon dioxide. According to the 2013 IPCC report, in the first two decades of its release into the atmosphere, methane is 86 times more powerful than carbon dioxide in terms of global warming force. Even over the course of a century it is 34 times more powerful.[61; 81; 82]

Methane is being released from Arctic permafrost at an increasing rate due to rapidly warming temperatures at the higher elevations. This photo shows a person standing on a frozen lake in Siberia. She made a hole in the ice and ignited the methane. (Photo, courtesy of Katey Walter Anthony's research team, University of Alaska Fairbanks. This research was funded by the National Science Foundation.)

The methane "mega-fart," mentioned in Chapter Three, caused "sudden" warming and subsequent mass extinction 55 million years ago.[83] But even that event occurred over centuries.

Two well-publicized sources of methane include methane escapes from drilling for petroleum and natural gas, and cows. The biggest store of methane, however, is underground in the permafrost and in the sea bed. As the earth now warms, methane is starting to be released from the permafrost. There are lakes in Europe and Asia where so much escaping methane has been trapped under the ice in winter that people can drill holes in the ice and ignite it. A few enterprising folks have even used the gas to heat their homes.

Methane clathrates, or hydrates, are methane in the form of slushy ice in the sea floor. They form when methane gas escapes from the earth and hits near-freezing deep water. It is estimated that there is more methane in this form than there is petroleum in the world. As the ocean warms, this methane will eventually be released, like tiny bubbles in soda water, with the result of dramatically increasing the greenhouse effect.

Such a release of methane would occur over vast areas of open ocean. To date there is no practical way to control or contain methane in that form. In short, while we do not know how soon the methane will be a big problem, the methane now coming out of the permafrost should be a signal for concern.[84]

The potential for a substantial methane release occurring is the subject of much scientific study. Recent work focuses concern on the outgassing of methane from permafrost in the Arctic.[85, 86] Since it is not possible to predict just how much methane will be released or when, scientists have not even included this possibility in their projections of climate change. While some predict that methane clathrates might not be released for a century or two, the unprecedented rate of warming could mean a much earlier release. The fact that such an event is not incorporated into either the climate models or sea level rise projections this century is another reason that the forecasts may underestimate, and are therefore conservative.

Underestimates In Coastal Elevation Data

A surprising source of confusion about the potential impact of sea level rise is the inaccuracy of traditional topographic maps and elevation data. Obviously, knowing the current height of a particular property above sea level is an important factor in assessing its vulnerability. One would think that land elevations are known and indisputable, but many elevation and flood maps are extremely inaccurate. In fact, many people live under a false sense of security from flood maps that indicate they are significantly higher above sea level than is true.

In the U.S., the usual authority on this topic is the National Elevation Dataset, maintained by the U.S. Geological Survey. Until recently, elevation mapping was done using rather simple surveying technology and a few data points. Although Global Positioning Systems (GPS) have greatly enhanced horizontal data references in the last few decades, they are of limited value in distinguishing elevations.

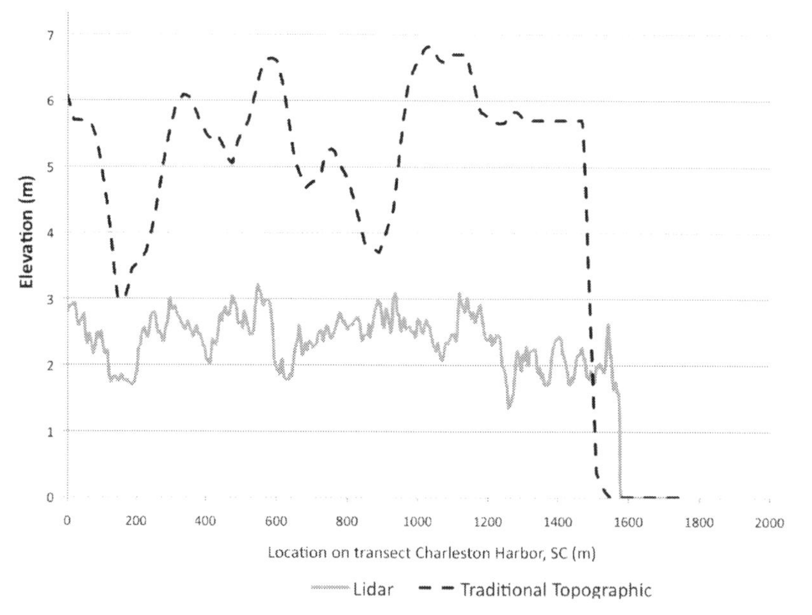

Figure 10-4. New LIDAR mapping is far more accurate than the traditional topographic maps that are still frequently in use. In this transect from a Charleston, South Carolina, harbor, the elevations were found to be as much as 15 feet lower than the historic flood maps showed. (Graphic, courtesy of NOAA-CSC.)

Fortunately, new laser-based LIDAR (Light Image Detection and Range) mapping, done from airplanes, is gradually replacing the old maps with dramatically improved resolution to within inches in accuracy.

Charleston, South Carolina, is just one of thousands of vulnerable coastal cities. Figure 10-4 compares the traditional elevation maps with the new LIDAR-based maps done in 2007. [87]

In this one example, true elevations are as much as 10-15 feet (three to four meters) lower than the topographic maps showed. That has tremendous implications for the impact of sea level rise, storm damage, and erosion. In the U.S. and internationally, multiple government agencies are working with private companies to update maps using consistent LIDAR data. The estimate is that 60 percent of the United States is presently mapped, though not to a consistent standard.

A recent study done for the federal government showed that 224 million dollars a year for five years would fund LIDAR mapping data for the whole country.[88] Even though they documented a benefit-cost ratio of at least four to one, and potentially 40 to one, this effort is not fully funded. At a cost of less than a dollar per year per taxpayer and a clear return on investment it would seem like an easy decision. Hopefully, the work will get done soon as it has great value to property owners, the insurance industry, the financial sector, and for disaster preparedness.

Population Impacts Very Vague

It is also very challenging to estimate the property and population that will be impacted by sea level rise. One reason for this is that until recently, population was not mapped with concern to accurate coastal elevation. So, when someone tries to estimate how many hundreds of millions of people will be displaced by sea level rise of one, two or three meters, the data is somewhat lacking. As a result you will see considerable variation in the estimates.

Furthermore, the impacts of sea level rise are not limited to elevation. Erosion and storm surge will compound the effects of a changing shoreline.

Thus, it is very difficult to assess how much land, how many homes and businesses will be destroyed. Beyond that, we cannot know how economic values of a given coastal location will be affected.

If rising sea level eliminates those homes and businesses, it is not necessarily true that they will simply be able to relocate on higher ground. There might be geographic obstacles to relocating, or it may simply not make sense to reestablish a marine community knowing that sea level will keep rising for centuries. The inevitability of long-term rise will change our cultural attitudes about where to invest.

Whether that community is tourism-based, a marina community for recreational boating, fishing, scuba diving, or a commercial port, the elimination of coastal real estate will have far greater implications than the value of the property that will disappear.

Perhaps one reason for the underestimated projections of scientists is the same reason the issue is confused for the rest of us—that it creates cognitive dissonance, and leads to the very human tendency to want to be happy and think good thoughts. In a world filled with challenges, there are only so many problems that we can embrace. Who among us has never felt that they don't want any more bad news?

Thinking deeply about this means thinking about the demise of our entire coastal civilization, possibly our own lifestyles and personal assets. It is hard to avoid the question: how bad could it get? Could all the ice melt in a thousand years, or even hundreds, causing that 212 feet or more of sea level rise?

I decided to put the question squarely to the expert, Dr. James Hansen. His reply, with my clarifications in square brackets: "The rapidity of the business as usual human-made climate forcing, burning all the fossil fuels in the next century or two, has no paleoclimate analog [ancient climate record to study as precedent]. With such a forcing, I would expect the time scale for demise of the great ice sheets [Antarctica and Greenland] would be measured in centuries, not millennia."

After I got that e-mail I realized that I too was subject to a form of cognitive dissonance. I had trouble getting my mind to accept that Hansen was saying that on our present path the meltdown of the ice sheets within centuries was a probability, not just a possibility. Every time I honestly think about that forecast, my mind quickly veers off, simply wanting to avoid the enormity of it.

Politicians, the press and the public seem to operate from the same inclination towards denial, procrastination, and preference to "live in the now" that can be seen in our policies and attitudes about national debt, social security, health care, and pension plans.

A *conservative* viewpoint on most issues tends to consider the long-term perspective. With sea level rise, it seems to me that a truly conservative attitude would be that there is no point just hoping things will get better, and certainly no point in denying reality. The facts are the facts. Not only in the United States, but globally, we have enormous capacity for innovation and adaptation. We should be positive, based on realism, not some Pollyanna view of things.

Famous historian Will Durant said, "Civilization exists by geological consent, subject to change without notice." In the case of the impending sea level change we do have notice. That knowledge gives us the privilege to plan and to adapt.

The Impacts

HOUSE FOR SALE

"Fisherman's Dream" 4 Bedroom, 3 Bath
Completely renovated. Historically desirable community. Could be oceanfront soon. Good fishing several times each month from combination driveway/boat ramp.
~~$599,000.~~ Reduced to $299,000
Act now. House will not last long.
Contact: clear-hindsight@net.com

Chapter Eleven
A Few Feet–What's the Big Deal?

Sea level rise may be the ultimate slow emergency, the kind of problem that invites procrastination. In most countries today, there is a strong tendency to avoid dealing with longer-term problems, whether they be economic, social, or environmental. In fact, crises in these three categories seem to have come to a head simultaneously, creating the oft-cited "perfect storm" that will worsen the impact of each.

To understand the magnitude of sea level rise, we have to consider the current rate of rise, the rate we can expect over the next several decades, and projected rates for the end of the century and beyond, based on different greenhouse gas scenarios.

Presently, sea level is rising at a rate of about an inch a decade. This rate has doubled over the past three decades. Lest we are tempted to think that inches don't matter, when Hurricane Irene barreled up the East Coast of the U.S. in August 2011, there was real fear that it would flood lower Manhattan. When it had passed, headlines proclaimed that just one more inch of sea level rise could have flooded subways and tunnels there.[89]

Just one more inch can lead to the collapse of dikes, levees, and seawalls. Once water starts to trickle over an earthen wall, the flowing stream of water can destroy the entire berm within minutes. The same can even happen with concrete dams and seawalls, though not as quickly.

As covered, most projections presently are for between three and seven feet of rise by the end of the century. Let's take a look at what this means.

A rise of just a few feet means different things in different places, depending on the topography and composition of a particular shoreline. A vertical granite cliff will be minimally affected. In contrast, a beach sand dune, with a lower area of marshland behind it, could lose more than a mile of property for each foot of sea level rise.

Just as an example, Figure 11-1 features a map section of Broward County in southern Florida. The solid black coloring indicates the areas that will be underwater at seven feet (two meters) of sea level rise. Even though such a situation is likely a century away, it helps to see the long-term outcome. With such extreme devastation, it is very questionable if even the remaining high-elevation areas would be viable places to live.

Section of Broward County Florida

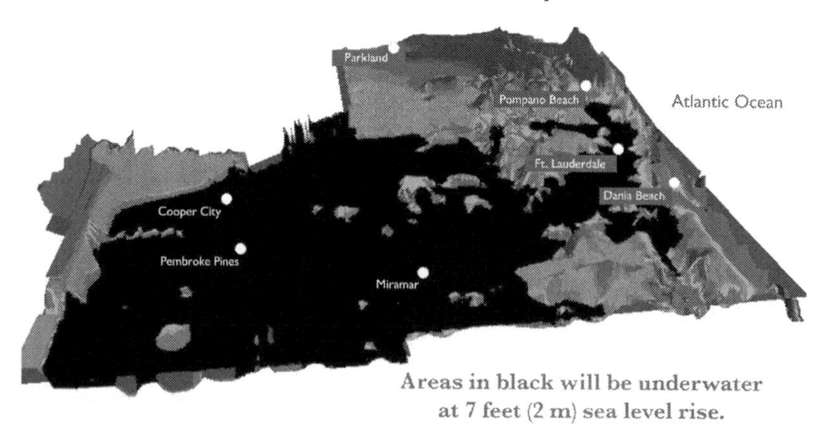

Figure 11-1. An image of Broward County, Florida, illustrating the area that will be underwater with a seven foot (two meter) rise in sea level. Surprisingly, most of the inundated area is a considerable distance from the ocean. (Photo, courtesy ESRI / Dr. Lin Wu, CSPU- Pomona.)

Frankly, I doubt that South Florida could maintain its current vibrant community once sea level rises beyond three or four feet. On the current path, this scenario will likely occur within a century. In 2011, the four southeast counties—Palm Beach, Broward, Miami-Dade, and Monroe (Florida Keys)—published a consensus outlook study on projected sea level rise.[90] Based largely on work by the U.S. Army Corps of Engineers, they have accepted a probable forecast of three to seven inches of sea level rise by 2030 and nine to 24 inches of rise by 2060.

There are some very useful models now available on the internet that provide visualizations of what various amounts of sea level rise will mean to the coastline. Three of the best are: surging seas http://sealevel.climatecentral.org/surgingseas/ by Climate Central. A similar tool is available at http://www.csc.noaa.gov/slr/viewer/. More tools and information are available at Digital Coast http://www.csc.noaa.gov/digitalcoast/ by NOAA's Coastal Services Center.

Stormy Weather

Until recently many climatologists and meteorologists questioned the correlation of the increased warming this past century with increased storm activity. That is rapidly changing. Though the correllation is not direct, most climate researchers believe that increased global temperatures cause a greater number of storms, including powerful hurricanes.[91] Heat is the driving force. Hurricanes spawn in locations of great ocean warmth, during the hottest months of the year. Figure 11-2 shows the increase of major hurricanes over the last few decades, the same period of record high atmosphere and ocean temperatures.

The National Hurricane Center analysis of hurricane damage for the U.S. recognizes a trend of increasing damage. Their 2011 study covering 1900-2010 found that nine of the 10 most damaging hurricanes occurred in the last decade. Even adjusting the costs for inflation, six out of the 10 most expensive were in that decade.[92]

As sea level rises, every storm has greater potential for damage. As Ben Strauss from Climate Central puts it, "You can think of raising sea level as being similar to raising the floor of a basketball court. It's easy to see that the point scores will change dramatically."

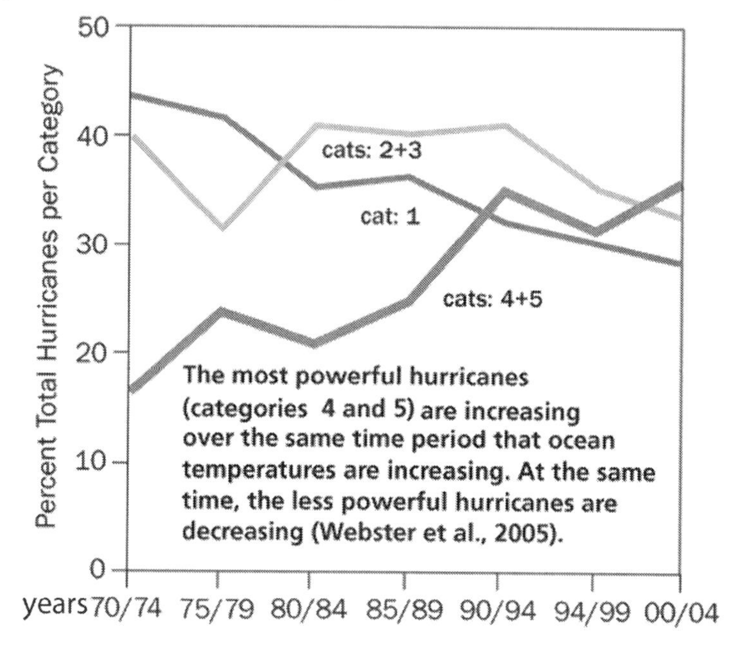

Figure 11-2. Recent history shows stronger storms. This is in agreement with computer models predicting more powerful storms with rising temperatures.[77] *(Graphic, courtesy of the Union of Concerned Scientists, www.ucsusa.org. Data from Webster et al. 2005.)*

Coastal real estate will increasingly be destroyed by surging storms, and literally hundreds of millions of people will be displaced this century. Hurricane Katrina gave the U.S. a sense of what suddenly relocating a million people looks like. While the overall dislocation will happen over years and decades, there will at times be sudden displacements when major storms devastate coastal communities.

Coastal communities where land is subsiding are likely to be affected sooner than others. Examples include New Orleans, Venice (Italy), and Bangkok (Thailand).

A study by the U.S. Geological Survey shows that the New Orleans area is subsiding at approximately 0.2 inches (five millimeters) per year, which means that the effective rate of sea level rise there is about 2.5 times greater than the global norm.[94]

Harbors And Major Infrastructure

As coasts are destroyed, there will be major infrastructure losses, including ports, coastal airports, roads, railways, military bases, sewage treatment plants, and more.

Just to take one example, seaports will be seriously compromised as sea level rises. Currently, facilities to load and unload vessels of all sizes are engineered to accommodate the movement from high tide to low tide, allowing for extremes of the monthly lunar cycle as well as historic ranges of storm surge.

Although there are a few places that have unusually large or small change, a typical tide range is about three feet from highest to lowest points.

Absolutely massive amounts of concrete and other infrastructure are all designed around that given range of tidal movement, in ports from Oakland to Singapore, as well as the small ancient fishing harbors in the Mediterranean, and on coasts around the world. Trillions of dollars of this one type of infrastructure and equipment will be severely compromised.

In 2007, the Organization for Economic Cooperation and Development (OECD) published a farsighted study of the port infrastructure problem, "Ranking Port Cities with High Exposure and Vulnerability to Climate Extremes".[95] It ranked 136 major port cities, internationally, for current impact as well as that projected for the 2070's. Figure 11-3 shows the 20 port cities in the world with the most exposed assets based on current levels of population and commerce.

Looking ahead to the 2070 scenario, the study projects damage totaling 35 trillion dollars in the 136 cities. To put this in perspective, that is more than twice the value of the Gross Domestic Product of the United States in 2010.

Beyond ports and marinas, major infrastructure located within modest reach of high tide and storm surge exists in innumerable coastal communities throughout the world. Some examples are: power plants, electrical distribution systems, sewage treatment plants, coastal airports, roads, railroads, refineries, fuel storage and transfer facilities, telecommunications networks, emergency response facilities and centers, and general civic infrastructure, from retail to landfills.

Each of these is vulnerable to the steady creep of rising sea level, extreme tides, and unpredictable storm surge, representing huge challenges and cascading effects from their potential incapacitation.

When the tsunami knocked out Japan's Fukushima nuclear power plant, we got another glimpse of the tremendous repercussions from unexpected damage to key utilities and infrastructure. That disaster was a direct hit on just one critical location. Rising sea level will not be selective. It will hit the developed and undeveloped worlds without prejudice.

MORE COSTLY SURPRISES

There are endless examples of impacts that you would likely never consider. Imagine, for example, the damage to the South Florida Water Management District, a vast system that controls water levels from Orlando to the Florida Keys.

A 2009 study showed that just eight inches of rise in ocean level would significantly compromise 28 gates and dams used to control everything from groundwater to flooding. Many of these gates were built in the middle of the last century, when there was no expectation of rising sea level, and gravity moved the water. Higher sea levels mean converting to massive pumping stations.

The study showed that 18 gates would have to be rebuilt for a sea level rise of just eight inches (20 centimeters). It would cost an estimated 70 million dollars to replace each with a new pumping facility, for a total of 1.25 billion dollars—just from eight inches of rise. Again, this is just one example to illustrate how hard it is to identify and estimate the costs of the consequences of even moderately rising sea level.[96]

Present Day Top 20 Port Cities by Asset Exposure

City	Country	Official 2005 Total Population ('000)	Current Population Exposed to Climate Change Risks ('000)	Current Assets Exposed to Climate Change Risks (US $ Bn)
Miami	USA	5,434	2,003	416
New York-Newark	USA	18,718	1,540	320
New Orleans	USA	1,010	1,124	233
Osaka-Kobe	Japan	11,268	1,373	215
Tokyo	Japan	35,197	1,110	174
Amsterdam	Netherlands	1,147	839	128
Rotterdam	Netherlands	1,101	752	115
Nagoya	Japan	3,179	696	109
Tampa-St Petersburg	USA	2,252	415	86
Hampton Roads, VA	USA	1,460	407	84
Guangzhou Guangdong	China	8,425	2,718	84
Boston	USA	4,361	370	77
Shanghai	China	14,503	2,353	73
London	UK	8,505	397	60
Vancouver	Canada	2,188	320	55
Fukuoka-Kitakyushu	Japan	2,800	307	48
Mumbai(Bombay)	India	18,196	2,787	46
Hamburg	Germany	1,740	261	39
Bangkok	Thailand	6,593	907	39
Hong Kong	China SAR	7,041	223	36

Source: Nicholls, OECD 2007

Figure 11-3. A list of the top 20 ports in terms of exposure to sea level rise. The same study also projected growth and expansion over the next 60 years. The ranking and assessment of exposure was considerably different in that future risk assessment.

FRESHWATER SUPPLIES

One of the most common sources of life-sustaining fresh water is "groundwater", drawn from underground aquifers or streams. In coastal areas, freshwater aquifers often lie on top of salt water, which is driven underground by the pressure of the ocean. As sea level rises, salt water will intrude on the freshwater supply and eventually destroy coastal freshwater aquifers. This situation is already occurring, as was reported in the Fort Lauderdale *Sun-Sentinel* newspaper in September, 2011.[97]

Once a freshwater supply is destroyed, the replacement costs will become a challenge for almost any community. A recent study showed that just three inches of sea level rise will intrude on the South Florida freshwater supply and also impede the control of storm water.[98]

CORAL REEFS AND OCEAN ACIDIFICATION

Hundreds of millions of people all over the world live in communities that are sustained by coral reefs. Some depend on the physical protection that corals provide; many more are supported by the tourism connected to those gorgeous, vibrant, and colorful "cities under the sea". In some cases reefs also sustain fisheries for food or for the lucrative aquarium business.

The threat to coral reefs does not come directly from sea level rise, but from a tangential effect. As explained, much of the carbon dioxide spewed into the atmosphere is absorbed by the ocean. The result is that the ocean turns more acidic, or, to be more precise, less alkaline, on the pH scale.

Over the last two centuries, ocean acidity has increased by a stunning 30 percent.[99, 100] The acidity level, measured as pH, has a big effect on the balance of life in the ocean.

Shellfish and corals need a calcium-rich environment to live. If the pH gets too low, making the environment acidic, shellfish and corals cannot survive. Some areas are already experiencing the economic consequences of this. For example, the billion-dollar Pacific Northwest oyster farming industry is struggling to keep its product from disintegrating from acidification.[101] Oysters are only one shellfish product. All of them are at risk from the increasing acidification.

Damage to coral reefs is another example of the impacts that climate change may have on the sea. Throughout the planet's history, the combination of ocean acidification and rising temperature have wiped out coral reefs at least five times.[102] The increasing pace of reef destruction over the last few decades is unprecedented and is already affecting communities where coral reefs are central to the economy.

I have personally witnessed this decline. For about 25 years I was in the resort scuba diving industry and traveled all over the Bahamas, the Caribbean, and occasionally to destinations in the South Pacific. I vividly remember the massive Elkhorn corals as a dominant feature of the vibrant coral reefs. They have now almost disappeared from most areas.

Colleagues in the diving industry and scientists from other parts of the globe quietly talked for several decades about coral die-offs, even in the most remote regions of the planet. Twenty-five years ago, I saw a few areas of bleached coral in the remote Maldive Islands in the Indian Ocean. A friend recently dove there and told me that coral bleaching is now widespread.

In Australia, the Great Barrier Reef has importance well beyond tourism. To prevent the destruction of this massive asset, the country is even experimenting with sunscreen canopies to slow the die-off. Given that the reef is more than 100,000 square miles in size, it is very doubtful that this is a practical, large-scale solution.

In late 2011, Dr. Peter Sale published a book with the worrisome title, *Our Dying Planet: An Ecologist's View of the Crisis We Face.* Sale looks at some of the ways a warming planet will affect our ecosystems. He does hold out some hope and prescribes how we might turn things around, if we act soon. Failing that, one of his startling conclusions is, "Coral reefs are on course to become the first ecosystem that human activity will eliminate entirely from the earth."[103]

Against that ominous forecast, here is one example of inventiveness. Dr. Brian von Herzen founded The Climate Foundation, a nonprofit organization that looks for innovative ways to deal with the climate crisis. Brian and his colleagues have developed a technique to use extremely simple "pumps", powered only by wave action, to lift cooler water from the depths and cool coral reefs during the hottest times of the year. Use of the pumps at a few experimental sites has shown remarkable success. (Link to more information at johnenglander.net/oceanpumps.) Though fascinating and inventive, we cannot rely on this technology to protect a significant portion of the world's coral reefs.

National Security Impact

Governments have started to assess the impact of rising sea level and other aspects of climate change at various levels. Military and intelligence agencies are at the forefront of this type of research and planning. Several documents provide a glimpse at how seriously they take the issues in terms of national security.

Every four years the U.S. Department of Defense must create and publish the *Quadrennial Defense Review*. The following are a few excerpts from the February 2010 document:[104]

- *Other powerful trends are likely to add complexity to the security environment. Rising demand for resources, rapid urbanization of littoral regions [close to shore], the effects of climate change, the emergence of new strains of disease, and profound cultural and demographic tensions in several regions are just some of the trends whose complex interplay may spark or exacerbate future conflicts.*

- *...in 2009, that climate-related changes are already being observed in every region of the world, including the United States and its coastal waters. Among these physical changes are increases in heavy downpours, rising temperature and sea level, rapidly retreating glaciers, thawing permafrost, lengthening growing seasons, lengthening ice-free seasons in the oceans and on lakes and rivers, earlier snowmelt, and alterations in river flows.*

- *In 2008, the National Intelligence Council judged that more than 30 U.S. military installations were already facing elevated levels of risk from rising sea levels.*

In 2007, CNA, a Washington D.C.-area think tank that works extensively on defense issues, assembled a group of senior, retired military leaders to examine the issue. Their report, "National Security and the Threat of Climate Change", is available online.[105] They evaluated issues ranging from the elimination of low-elevation bases such as Diego Garcia in the Pacific to the tens of millions of refugees displaced by the rising ocean. To quote from their Executive Summary:

> *The predicted effects of climate change over the coming decades include extreme weather events, drought, flooding, sea level rise, retreating glaciers, habitat shifts, and the increased spread of life-threatening diseases. These conditions have the potential to disrupt our way of life and to force changes in the way we keep ourselves safe and secure. In the national and international security environment, climate change threatens to add new hostile and stressing factors. On the simplest level, it has the potential to create sustained natural and humanitarian disasters on a scale far beyond those we see today. The consequences will likely foster political instability where societal demands exceed the capacity of governments to cope.*

While some might recall cases of poor military planning, the military might actually be in one of the best societal positions to understand and prepare for sea level rise. As an example, Admiral David Titley served as oceanographer of the U.S. Navy until 2012 and was one of those taking the lead on this issue. He has an advanced degree in oceanography as well as meteorology and climatology, and led the Navy's Task Force on Climate Change. Titley admits that he was skeptical of climate change and the connection to greenhouse gases not too many years ago. By the time he became the Navy's lead person on the issue, he saw a new reality and assertively explained the science to Pentagon and other officials.

Having heard him numerous times and been on panels with him, I find Dave a particularly convincing speaker. Explaining the many challenges the Navy will face, not the least of which is damage to facilities, he wryly observes, "While the ships may float, for some reason we built all our navy bases at sea level."

In one final view of national security aspects, consider a third report with exceptional scope and an extensive array of impressive authors, "The Age of Consequences: The Foreign Policy National Security Implications of Global Climate Change'.[50] This report looked at three plausible scenarios: Expected, Severe, and Catastrophic.

Former CIA Director James Woolsey wrote the chapter on the third scenario. To briefly excerpt:

> *In a world that sees two meters of sea level rise, with continued flooding ahead, it will take extraordinary effort for the United States, or indeed any country, to look beyond its own salvation. All of the ways in which human beings have dealt with natural disasters in the past... could come together in one conflagration; rage at government's inability to deal with the abrupt and unpredictable crises; religious fervor, perhaps even a dramatic rise in millennial "end of days" cults; hostility and violence toward migrants and minority groups at a time of demographic change and increased global migration, and intra- and interstate conflict over resources, particularly food and fresh water. Altruism and generosity would likely be blunted.*

Note that the catastrophic scenario he addresses comes from a two meter (seven foot) rise in sea level, a projection currently on the high side, for this century. Whether that timetable speeds up, or does not occur until the early decades of next century, will not greatly change his assessment.

It is some comfort to know that at least the military and national security agencies are starting to plan for the inevitability of this long-term emergency that will have essentially permanent consequences.

Chapter Twelve
Miami, Maldives and Much More

Tiny Rhode Island is known as "the ocean state" for its high proportion of shoreline. Years ago, then-Senator Ted Green was needled by a colleague, who asked, "Senator, just how small is that little state of yours?" Senator Green paused, then responded, "Well, that depends: high tide or low tide?"

In the decades ahead, that quip will take on new meaning, and not just in Rhode Island. Let's take a look at a few locations that demonstrate the diversity of sea level rise impacts, keeping in mind that almost any coastal city in the world could be highlighted.

Miami, Florida

Miami is used as the "poster child" of potential sea level rise impact, for good reason. Greater Miami is a stunning city, from the famous art deco district of chic South Beach to the hundreds of high-rise office buildings and thousands of million-dollar condominiums. It's a vibrant city, a cultural crossroads, a tourism Mecca, and the business hub of the Caribbean and Latin America.

When it comes to sea level rise, it is also one of the most vulnerable cities in the world. In his provoking and well-researched book, *The*

Flooded Earth, Dr. Peter Ward opens with a futuristic description of Miami in the year 2120, with sea level 10 feet higher than at present. Little remains of the current metropolis, and the crumbling Miami skyscrapers are protruding from the water. While this scenario is slightly on the high side of most estimates, it is very plausible. Even before that amount of sea level rise occurs, Miami is in big trouble. The average elevation of Dade County is just four feet (1.2 meters) above sea level.[106]

In July 2013 a feature story appeared in Rolling Stone Magazine titled "Goodbye Miami" that sent shock waves through the community. With extensive research it described a futuristic scenario that 2030 was the beginning of the end.

Miami's civic leaders are very aware of the issue and have established a Climate Change Advisory Task Force. For practical reasons, they limited their primary focus to the year 2060, just five decades ahead. As mentioned earlier, their most recent assessment predicts sea level to reach nine to 24 inches by then.[90] This estimate may be low, but even if accurate, the results are severe.

Already, Miami and the Florida Keys have spent millions of dollars to raise the elevation of a few streets by one or two feet because they were awash at extreme tides. The trend is clear. This is only the beginning.

While there is increasing awareness of South Florida's vulnerability, many assume the problem can be solved with seawalls and beach replenishment. Such beliefs ignore the region's geologic Achilles heel, the porous limestone base that I will address shortly.

The sad truth is that Miami will eventually become the northernmost island in the Florida Keys, like nearby Key Biscayne. Local governments need to make changes to zoning, building requirements, and long term planning recognizing the reality of sea level rise. The sooner the better. Even with the use of elevated buildings, the impact on the huge metropolitan area will be devastating. Elevating entire sections of the city is possible but extremely expensive.

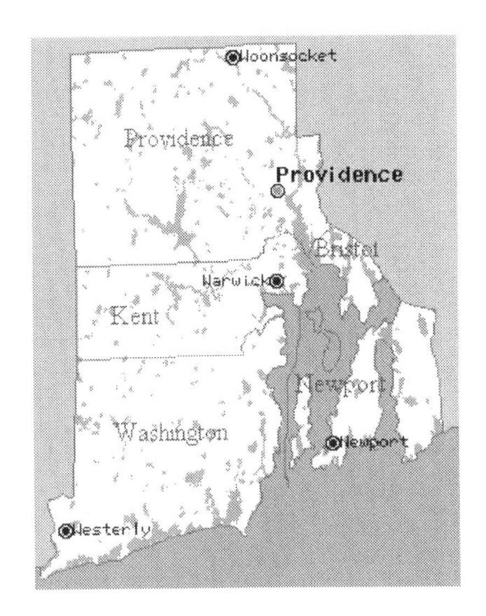

Figure 12-1. The funnel shape of Narragansett Bay creates a greatly exaggerated exposure to sea level rise for the state capital of Providence, Rhode Island, and nearby communities. At high tide and with certain wind conditions the water effectively piles up, amplifying the ocean height. This adds to their problems from rising sea level. It is a good example of how geologic and topographic features need to be considered for specific adaptation plans. (Image from Wiki Commons.)

PROVIDENCE, RHODE ISLAND

Returning to Rhode Island, the current Senator, Sheldon Whitehouse, is very concerned with sea level rise, pointing out that Providence and the surrounding area are extremely vulnerable. Aside from the typical exposure of a low-lying coastal city, this area has a unique feature that amplifies the impact as sea levels rise.

The state capital is located at the northern end of the rather large, triangular-shaped Narragansett Bay. Certain wind conditions funnel water into this area, greatly amplifying high tide. A global sea level rise of less than a foot will produce significant problems, worsening those already being experienced in the state capital. In addition, the area has been subsiding, approximately six inches over the last century. (See johnenglander.net/providence).

BOSTON, MASSACHUSETTS

One of America's oldest cities, Boston is particularly vulnerable to sea level rise because of its low elevation and the fact that it was built on landfill. Islands at the mouth of the harbor also act as a funnel during certain storms. The large arm of Cape Cod to the south increases this problem.

A 2009 study by Allianz Insurance and World Wide Fund for Nature calculated that an 18 inch (45 cm) rise in sea level by 2050 would put over 400 billion dollars of Boston area assets at risk.[107]

Boston is one area where the local government and the corporate world are starting to look at the issue and options. As covered in more detail in chapter 16, plans have been suggested for a bold, creative, longer-term solution to issues of sea level rise, erosion, and storm surge.

On select projects Boston has already embraced the prospect of higher sea levels. The new wastewater treatment plant on Deer Island allows for a three foot rise in sea level. Unfortunately the larger, long-term plans still remain on the shelf.

NEW ORLEANS, LOUISIANA

When Hurricane Katrina hit this low-lying city in 2005, water breached levees in more than 50 locations and caused massive devastation to homes and infrastructure. We will avoid retracing that well-known situation, except to point out that New Orleans and vicinity face a triple threat:

- The land here has been subsiding as a result of the compacting soils and the removal of water and petroleum.
- The city is already below sea level and depends on a network of levees that is already inadequate for storm surges. These levees cannot be heightened much because of the poor soil upon which they are already built.
- The city is located in a frequent hurricane path.

NEW YORK CITY

[Editor's Note: This section is unchanged for this second edition, appearing just as it was first published 10-22-12; Hurricane Sandy hit Atlantic City and New York City one week later.]

Manhattan, the iconic borough of New York City, is often thought of as being highly vulnerable to sea level rise. While there is obviously some truth to that, let's put it in perspective. It is not a sandy beachfront island, and the average elevation is 54 feet,[108] with some areas exceeding 250 feet above sea level.[109]

Lower Manhattan does have a considerable amount of real estate at low elevations, but even most of that has 10 feet of clearance to the high-tide mark. Equally important is that its geology of granite, gneiss, and other hard, impermeable rocks makes it quite defensible. These types of rocks will serve as excellent bases for the areas where seawalls will be needed for protection, as compared to South Florida and other coral limestone locations.

New York City does share one similar concern with Boston and Providence. The broad arm of Long Island, and the rivers around Manhattan that continue up the Hudson River Valley can, under certain conditions, act as a funnel, amplifying storm surge effects for Manhattan.

The *New York Times* carried an Op-Ed on September 25, 1999, entitled, "Hurricanes on the Hudson."[110] Perhaps few noticed it; many of those that did likely wanted to forget what they read. Erik Larson cited a 1995 study of what could occur if a category four hurricane approached New York City from a particular direction:

> *When researchers with the National Weather Service, working with the Army Corps [of Engineers], applied the model to New York City they discovered, to their great surprise, that the slope of the sea bed and the shape of the New York Bight, where the coasts of New York and New Jersey meet, could amplify a surge to a depth far greater than if the same surge had occurred elsewhere. The studies showed that a category four hurricane moving north-northwest at 40 to 60 miles an hour, and making landfall near Atlantic City—which would drive the storm's most powerful right flank into Manhattan—could create a storm surge of nearly 30 feet at the Brooklyn-Battery Tunnel. The water could rise as rapidly as 17 feet in one hour.*

Despite the massive potential damage, one can envision New York City being defended against 20 or 30 feet of sea level rise. Not only is it technically feasible, the enormous value and density of assets on that island makes it easy to rationalize a significant cost for the protection.

I don't want to minimize the challenge for New York City. Its residents and elected officials are legitimately concerned about the impact of sea level rise on tunnel and subway entrances. Also, the other four boroughs of the city have significantly lower elevations and greater exposures. Overall, however, Manhattan and New York are a very different situation from Miami, contrary to popular perception. Mayor Michael Bloomberg deserves real credit for identifying the problem of long-term sea level rise and implementing strategies to adapt.

Norfolk/Larchmont, Virginia

The metropolitan region encompassing Virginia Beach, Portsmouth, and Newport News is situated at the mouth of the Chesapeake Bay, the largest estuary in the United States. A natural harbor, it is home to Naval Station Norfolk, the world's largest navy base, including Norfolk Naval Shipyard.

This area, known as Hampton Roads, is already experiencing significant sea level rise. Measurements at Sewells Point show 14.5 inches of sea level rise in the last 80 years, just about double the national and global average. Oceanography Professor Larry Atkinson from Old Dominion University explained the exaggerated effect, "In addition to global sea level rise, we have two reasons for the faster rate locally: the last continental glaciation (18,000 years ago) was so heavy it pushed down the land in southern Canada and, like a teeter-totter, Virginia went up. Now, even long after the glaciers disappeared, southern Canada is rising and coastal Virginia is sinking. The second reason is that the lower Chesapeake Bay area is subsiding a little relating to the impact of a meteorite 35 million years ago, local compaction and groundwater withdrawal."

Retired Captain Joseph Bouchard, commander of the Norfolk Navy base from 2000 to 2003, has been very vocal in his concern about the threat of sea level rise for the base, which is already vulnerable to large storms and hurricanes. Bouchard said, "Parts of the base are actually below sea level, and significant parts flooded during Hurricane Isabel, a relatively weak hurricane." The Navy has already had to spend millions adapting to the higher sea level of the last few decades.

Navy bases and other port facilities have unusual problems with sea level rise. Often stretching for miles, they contain vast amounts of concrete and infrastructure that are intentionally designed to be as close as possible to sea level for easy access to ships. Moving or rebuilding such facilities is no small feat in the best of circumstances, never mind when sea level rise is predicted to continue for centuries. The situation presents a major challenge not only for Navy bases, but for all companies that are dependent on a global supply chain based in ocean freight.

Larchmont is a community immediately south of Naval Station Norfolk. That community's experience with rising sea level made the front page of the *New York Times* in 2010.[111] Some streets there were already experiencing two to three feet of flooding with every high tide, making driving a problem for as much as eight or nine days a month. The city finally agreed to raise one short stretch of road by 18 inches, at a cost of more than a million dollars.

Not only has high tide hit "main street" in this community, it has hit home. In the last few years, according to the same article, the Federal Emergency Management Agency (FEMA) paid to raise the elevation of six houses on the street, at a cost of 144,000 dollars each.

These communities are experiencing early effects of higher-than-typical sea level rise, offering a glimpse of what will happen to all coastal communities in the decades ahead.

SACRAMENTO, CALIFORNIA

FEMA, the U.S. Geological Survey (USGS), and the National Oceanic and Atmospheric Administration (NOAA) have evaluated communities most vulnerable to flooding related to sea level rise. Their surprising assessment was that the most vulnerable area in the United States in terms of levees may be in the vicinity of Sacramento, California.

Earthen levees span some 1,300 miles along the Sacramento-San Joaquin River Delta. The delta was reclaimed over a century ago, largely using simple techniques implemented by Chinese immigrant labor. It yielded some of the most valuable farmland in the world. Earthquakes pose an immediate risk in that they could weaken the levees, causing major flooding. Geologists and engineers figure there's better than a 60 percent chance of that event in the next quarter century. To cure just that one structural weakness is estimated to cost 700 million dollars, a daunting amount given the state's fiscal crisis.

But that fix would not even begin to deal with the damage from rising sea level. A seawater breach, whether caused by earthquake, storm surge, or gradually rising waters, will not only displace the local community, but jeopardize the drinking water for three-quarters of the state, and ruin some of the most productive farmland in the world.

It would likely take more money to fix this levee system than the one around New Orleans. As a reference for the cost and vulnerability of these levees, consider that in 2004 there was a breach of one levee not too far from Sacramento. It took six months and 90 million dollars to rectify. It is believed that the cause was damage by a beaver.[112]

San Francisco, California

The San Francisco Bay Conservation and Development Commission has looked at projected impacts on San Francisco Bay, including San Francisco and Oakland airports. See http://www.bcdc.ca.gov/planning/climate_change/climate_change.shtml. On that page, you can find a fascinating booklet, "San Francisco Bay: Preparing for the Next Level." Just to cite two facts from this publication: "There is general consensus that today's 100-year storm event may well be the normal water level by the year 2050, with a predicted sea level rise of 16 inches"; "The projected cost of adaptation, just for the San Francisco Bay area, is 62 billion dollars."

Their studies anticipate up to 55 inches of sea level rise this century, between four and five feet. Already, in San Francisco the statistical once-in-100-years flooding event is now a once-in-10-years event.

Seattle, Washington

Perhaps best identified as the world headquarters for Microsoft and Starbucks, the drizzly city has gained prominence over the last few decades. Seattle is not only a vibrant contributor to the U.S. economy, its port is a critical part of trade with the Far Pacific and Orient.

So far, its terrible exposure to sea level rise has been overlooked. Not only is Seattle vulnerable, but so is the entire area bordering Puget Sound. Anyone who has driven through the area will recall looking down at the Boeing factory and the expanse of buildings near the water. This is mostly reclaimed land just a few feet above the water. When it was developed in the latter nineteenth and early twentieth centuries there was no consideration given to sea level rise. A wooden seawall runs for hundreds of miles around the meandering shoreline.

Replacement of the aged and crumbling structure has been estimated at 250 million dollars. That would only replace the present wall, ignoring the rising ocean. Like nearly every other city mentioned, Seattle has huge budget imbalances. Even the short-term plan to repair the current structure has been put on hold.

I am not aware of any adaptation plans that would address the coming sea level rise in this area—rather remarkable when one considers how much Seattle exemplifies the modern environmentally-aware American techno-city.

WASHINGTON, D.C.

One U.S. city not on the coast is the capital, Washington, D.C., which is a good example of a critical city that is well inland, yet very vulnerable to sea level rise. The Potomac and Anacostia Rivers that come right through the city are navigable waterways connected to the sea.

Many areas of the city are very low-lying and already have flooded during extreme conditions. Historically, these areas were flood plains. The area just southeast of the White House is only four feet above sea level.

Over the coming century the city will experience some serious challenges (see Figure 12-2). Yet, Washington should be able to adapt to rising sea level better than most cities, at least for the next century. Its exposure is mentioned just as an example that the problem is not limited to locations on the coast. Any community on a river connecting to the ocean will face similar challenges.

Washington D.C. Flood Exposure Map

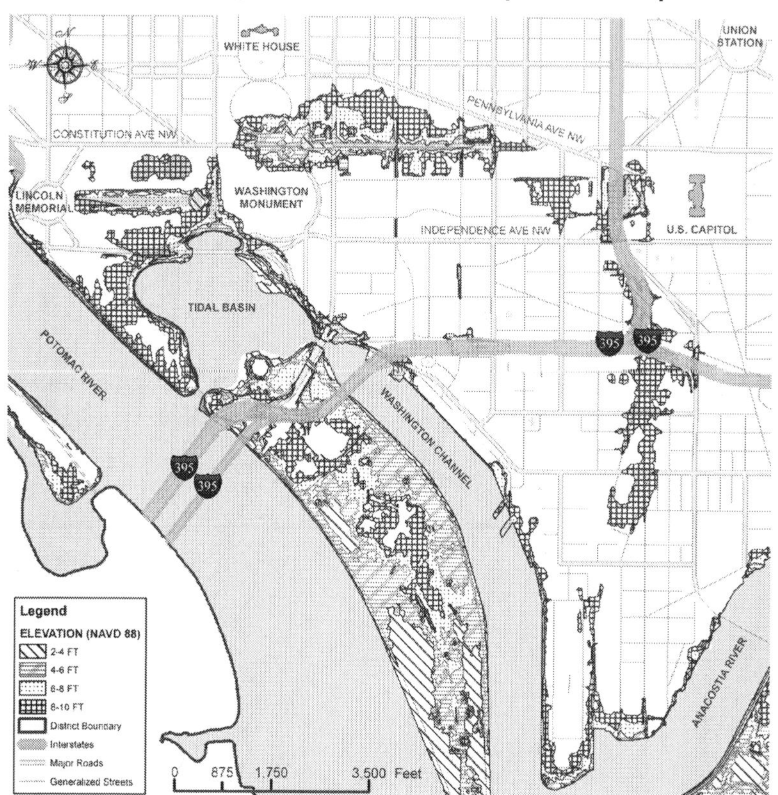

Figure 12-2. This flood exposure map shows the two navigable rivers, the Potomac and the Anacostia, connecting Washington, D.C., to the ocean, shown in light gray. The other shaded areas show low-lying land, some just four feet above sea level. For example, one is just blocks from the White House, at the top and slightly to the left. (Graphic, courtesy of D.C. Department of the Environment.)

International

A country's size, economic strength and natural resources will largely determine how well it will be able to cope with the sea level rise crisis. Aside from the glimpses of U.S. cities above, it is useful to look at impacts on a few nations.

Netherlands

The Dutch simply cannot avoid the dangers of its low-lying landscape. They have a long history of fear and reverence for the ocean. Many of us grew up hearing the fable of the Dutch boy who used his finger to plug a leak in the dike, saving a community. Those who live there are more likely to recall the reality in 1953, when as already mentioned, a strong North Sea storm killed more than 1,800 people during an overnight flood.

Of any country, Holland probably has the greatest historical experience with sea level rise planning and adaptation. In September 2008, the prestigious new Dutch Delta Commission reported that the Netherlands would need a massive new building program to strengthen the country's water defenses against the anticipated effects of global warming this century and next. The commission said the country must plan for a rise in the North Sea up to 4.25 feet (1.3 meters) by 2100, rather than the previously projected 30 inches (.8 meters), and plan for a 6.5 to 13-foot (2-4 meter) rise by 2200.[115] The plan includes more than 140 billion dollars in new spending this century.

SINGAPORE

Moving focus across the Pacific, the island-nation of Singapore is admired for its vibrant economy and its efficiency. This world hub might serve as an example of intelligent adaptation in a highly vulnerable situation.

Singapore is mainly low-lying, and much of its infrastructure is built on landfill. The business district, airport, and harbor are all at less than seven feet (two meters) above sea level.

Leaders have partnered with Dutch expertise to build a series of dikes and polders (protected areas behind the formidable dikes that are not brought up to sea level). This system is meant to provide sufficient protection for Singapore for the better part of this century.

As a side note, it is worth mentioning that small, progressive Singapore has taken a very proactive position on improving energy efficiency and reducing greenhouse gases.

MANILA, THE PHILIPPINES

Metro Manila consists of 16 cities with a total population estimated to be 21 million. It is ranked the 11th largest population center in the world today. In 2012, a study published in the professional journal *Natural Hazards* conducted by experts from the U.K. and the Netherlands listed Manila as one of the three most vulnerable cities in the world for flooding vulnerability and for economic impact of sea level rise.[113] Manila's vulnerability stems from its low elevation and seasonal flooding from multiple rivers carrying runoff from higher elevations. Record flooding in 2012 has heightened awareness about preparing for sea level rise. Though the city has plans to spend tens of billions of dollars to address the risk, many are concerned that these plans are grossly inadequate.

VENICE, ITALY

Venice is a very special city, perhaps analogous to New Orleans. It has history, architecture, and culture that cannot be relocated intact. While the same can be said for any community, these two have become national treasures, perhaps justifying extraordinary efforts and expense to preserve. Venice is the epitome of canal-based architecture. It is also the most vulnerable to all the threats from the ocean. When the city was established some 1,300 years ago, the astute engineers of the day realized that mud and silt do not make a good foundation. So, they drove hardwood piles into hard layers of compressed clay as a foundation for the city.

Then during the first half of the twentieth century Venice pumped significant quantities of water from deep underground, mainly for industrial uses. This caused the underlying substrate to compress and the region to sink. As a result, Venice saw a combined effective sea level rise last century of nine inches (23 centimeters), an inch or two more than the global average. By 1960 the problem was identified and the pumping practice was prohibited. Yet the basic problem continues.

St. Mark's Square is the most famous landmark in Venice. There, the pavement has been raised three times to cope with ever-increasing flooding. Even with that, it is now underwater approximately 100 days each year, a common situation locally referred to as Acqua Alta, or high water.

Flooding in the famous St. Mark's Square, Venice, Italy on September 18, 2009. For the last three decades, these have become increasingly routine, known locally as alta acqua, or high water. (Image from Wiki Commons.)

The Venetians are subject to a conflux of factors, including sea level rise, high volumes of runoff from the surrounding land during major rains, compacting substrate, and unusually high tides. A particularly disastrous and memorable day for them was November 4, 1966, when these factors convened to raise waters six feet, four inches (two meters) above normal.

In 2004, they initiated planning for the massive "MOSE" (Modulo Sperimentale Elettromeccanico) project, Italian for electromechanical experimental modules. The project, which is estimated to cost six billion dollars and has now been delayed until 2014, will result in 78 high-tech floodgates to seal off the Venice lagoon from the Mediterranean when conditions warrant.

One big complication is water circulation. This is not some esoteric oceanographic concern. The stench of garbage and sewage in the Venice lagoon can be, shall we say, breathtaking, on a warm day when the currents are not there to flush the canals. As the MOSE gates are raised more often and for longer duration, it will make the lack of flushing the canals a serious problem. The engineering and original construction of the city apparently make a modern wastewater treatment system infeasible.

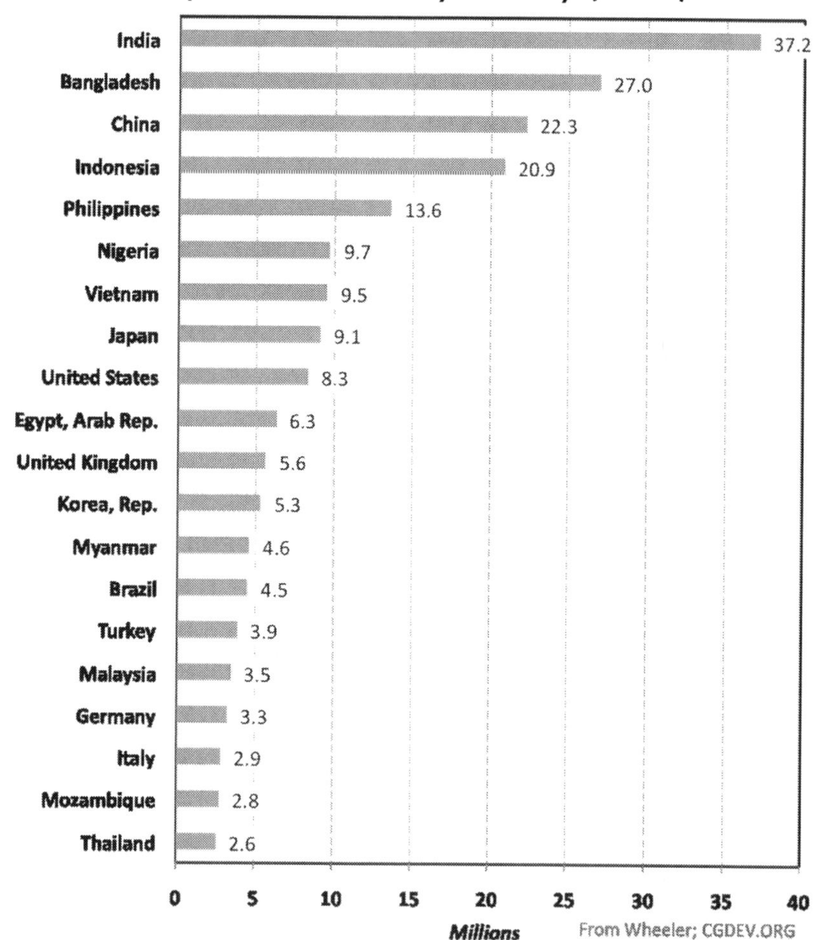

Figure 12-3. Rankings of exposure to sea level rise are completely dependent on the criteria and the projected timeframe. This graphic from the Center for Global Development (www.cgdev.org) shows the projected populations to be displaced by the year 2050. (David Wheeler, "Quantifying Vulnerability to Climate Change: Implications for Adaptation Assistance," Center for Global Development, 2011, Paper 240, page 21.)

Shanghai, China

The above-mentioned study also listed Shanghai as one of the three most vulnerable global cities. At 23 million people, it is now listed as the world's largest city. In addition to its vulnerable position at the mouth of the Yangtze River, Shanghai has subsided by eight feet (2.5 meters) over the last century. City leaders quickly denied that they should lead the "most threatened" list, citing the billions of dollars of defenses that they are constructing in an effort to protect the world's largest container port[114].

London

London, England exemplifies a city that is far from the sea but still at risk to its effects. Entwined by its relationship to the River Thames, over the centuries The City of London has transitioned from a tiny Roman outpost to one of the great global capitals in business, government and culture.

Metropolitan London covers more than 3,000 square miles and has a population of 12 million. It lies upon the Thames Valley, a floodplain that includes rolling hills. Now considered one of the leading financial and cultural centers of the world, the city itself is steeped in history, including four world heritage sites and many other instantly recognized landmarks.

In order to understand the impacts that sea level rise will have on London, one must closely examine the River Thames. The Thames is approximately 215 miles long starting in Gloucestershire UK and travels through central London before emptying into the North Sea. The Thames itself has also for a long time played a central role in London's economic dominance serving as both a military asset and as a highly lucrative port bringing in goods from all over the British Empire during its glory days. Today the Thames continues to serve as an important port, however most of its activity have been moved far downstream toward its mouth.

Before the Victorian era, the Thames River was far wider and shallower than what is seen today running through central London. Starting in the 19th century, efforts began to embank the river and reclaim some of the marshy land that had existed next to the Thames. Along with the embankments, the London sewer system was constructed at the same time and enclosed several small tributaries of the Thames in the London area and turned them into part of the sewer network.

It is important to note that the Thames experiences tidal action driven by the North Sea, where the river rises and falls as much as 24 feet twice a day, depending on the time of year. Because of that tidal connection, the Thames is also highly susceptible to storm surge, especially when the river is at high tide. For example, the North Sea Flood of 1953 when a high spring tide coupled with a strong storm surge to cause extensive flooding prompting the government to investigate ways to prevent recurrences.

The solution was the Thames Barrier, operational since 1982, that can be raised when necessary to prevent storm surge and extreme tide from flooding the city. With 119 flood defense closures by the start of 2010 it has been a great success. The Thames Barrier was designed to deal with a once-in-thousand year storm.

Rising sea level and subsiding land have now changed the outlook. It is estimated that problems will start to develop by about 2030. As of now there are no plans to replace the barrier with a larger one until 2070. If the rate of sea level rise exceeds expectations, that may have to change.

Fortunately by not being directly exposed to the ocean, London is in a very favored situation to engineer a solution that should be able to last a very long time.

MUMBAI

Mumbai, India provides a particularly strong example as to how a city built primarily on reclaimed land will react to rising sea level. Also known as Bombay, it is the most populous city of India as well as the fourth in the world with more than twenty million in the metropolitan area.

Mumbai originally consisted of seven islands inhabited by fishing colonies. Over the course of several centuries control of the islands passed through various indigenous empires. It was then taken over by Portuguese explorers followed by the British East India Company, the British Raj and ultimately India.

The seven islands were connected into one large island with the construction of the Hornby Vellard. This created a causeway uniting all seven islands around a deep natural harbor. Built two centuries ago, one of the goals was to stop the flooding of low-lying areas at high tide levels. That early massive engineering project positioned the city to become one of the most important in all of Asia.

Mumbai today is the commercial and entertainment capital of India, including Bollywood- India's movie production center. It has one of the highest concentrations of billionaires in the world. The opportunities that Mumbai offers attract many migrants from all over India hoping to achieve a higher standard of living.

The total area of Mumbai is approximately 233 square miles. Many parts of the city lie just above sea level though the average elevation ranges from 33 to 49 feet (10-15 m). Due to Mumbai's low elevation, as well as a foundation of mostly reclaimed sand, the city faces a risk of inundation by storm surges, as well as erosion, that will only increase as sea levels get higher.

In addition to compacting soils, the city has great seismic activity with 23 fault lines running through it. When looking ahead to rising sea level, this adds to the challenge of engineering sustainable structures.

One immediate impact is that the city will experience an increase in migration from other coastal regions, a daunting problem given the huge coastline and more than a billion people. For much of coastal India, fishing remains the lifeblood of the various communities. With increased sea level rise there has been concern about a negative impact on the low-lying small communities as well as the fisheries themselves. Many will likely migrate to Mumbai seeking opportunities, putting more strain on resources and potentially putting more people at risk for storm surges and extreme tide events.

TOKYO

Tokyo is Japan's largest city and capital and the world's largest metropolitan area. This business, banking and tourism center is home to approximately 35 million people, 51 of the Fortune 500 companies and has an annual productivity of 1.5 trillion dollars. Despite being such a valuable city, it is also one of the most vulnerable.

Originally known as Edo, the Japanese word for "estuary," the city partially sits on both a series of estuaries and a chain of islands. These low-lying areas make Tokyo rather susceptible to storm surges, tsunamis and ultimately sea level rise.

The mainland portion of the city is approximately 850 square miles, lies just northwest of Tokyo Bay and has an average elevation of 131 feet. The low land is near the coastlines and crisscrossed by various rivers as well as several chains of Islands.

Based on current projections, by 2070 sea level rise could directly impact two million Tokyo residents with several billion dollars in damages. However sea level rise will significantly magnify the current risks that Tokyo faces on a daily basis.

With the increase in sea level even at minimal levels, the impact of storm surges from typhoons (Pacific version of hurricanes) will magnify. Without added fortification, a storm of the magnitude of Hurricane Sandy towards the end of the century would have the potential to flood various parts of Tokyo causing damage that would easily surpass Sandy's by several times.

Being on a seismic hot spot, Tokyo is also at an elevated risk for powerful earthquakes and tsunamis. With rising sea levels, a tsunami would undoubtedly cause more damage to the city towards the end of the century than if it were to occur today.

Tokyo is not new to disaster; it has been destroyed and rebuilt twice due to both man-made and natural disasters. As a result of its destruction and rebirth, the city has more experience than most at mitigating disaster. For example, after the city was ruined by the great Kant earthquake of

1923, and by air raids in World War II, the city was rebuilt using state of the art building codes that mandate earthquake protected building as a way to prevent similar destruction.

For the fiscal year starting in 2013 the Japanese government plans to spend one trillion yen on flood mitigation. Tokyo by itself has set its sights on how to prevent disasters for the next two centuries. One of its primary goals is to prevent flooding along the city's many rivers due to heavy rainfall, storm surges and tsunamis. The plan is to build massive super levees along the rivers, to defend against a once-in-200 year storm.

In short Tokyo has experience with destruction and rebirth. The government appears to be preparing itself with steps to minimize the impact of sea level rise on this iconic city.

VANCOUVER

Vancouver, Canada is the business and cultural center of British Columbia. It is one of the world's most vibrant and dynamic cities benefiting from Canada's stability and natural resources, proximity to the US and a very strong Asian community. No modern metropolitan city has its combination of nearby snow-capped mountains, nearby islands, rich marine life, temperate climate, and manageable population level. All that desirability likely accounts for the sky-high real estate prices.

There is an invisible weak spot however. Many of the highly valuable properties are built on delta land at the mouth of the Fraser River. Like all deltas the silt that has been deposited over millions of years is prone to compaction, magnifying the impact of sea level rise. Parts of Metro Vancouver, including the airport, ferry terminals and industrial areas sit on sand deposits that have already displayed subsidence from the sheer weight of the infrastructure.

To their credit the city and provincial governments are looking at climate change adaptation strategies such as enhancements to the four decade-old dike system. In some areas flood control gates may be the answer. New city building codes can prohibit building new infrastructure in low-lying coastal areas or mandate new elevation standards for projects.

Improving city sewage and storm-water systems will make them able to function with higher sea level.

A recent government study also acknowledged that eventually "retreat" or the removal of all infrastructures may be the ultimate solution in certain areas. All of these strategies will not come without costs. According to the British Columbia Ministry of Forests, Land and Natural Resource Operations, over the next hundred years, these approaches are estimated to cost 9.5 billion dollars. Whether that's a good estimate or not depends on how quickly sea level rises.

DEVELOPING COUNTRIES

According to a study completed in 2007 by the World Bank, "The Impact of Sea Level Rise on Developing Countries," the top three casualty nations will be the Bahamas, Vietnam, and Egypt.[116] That particular study only evaluated 84 major countries. In a moment we will focus on some smaller island states as well.

In any such tabulation, the criteria used can change the list. Using risk parameters ranging from population, percentage of population, property, or gross domestic product, China, Suriname, India, Bangladesh, Nigeria, and Sri Lanka rank high on lists of vulnerability.

The effects of sea level rise are complex and not always obvious. For example, Argentina's population is not particularly vulnerable in terms of residential area, but its agricultural land is highly exposed to the rising sea.

Mexico, Tunisia and Indonesia all have significant low-lying, highly populated coastal areas. India and Bangladesh top the list in terms of millions of people at risk as shown in Figure 12-3.

THE ULTIMATE SACRIFICE

Finally, the evaluation of impacts of rising sea levels has a unique category: nations that will be eliminated entirely. For low-lying island nations, rising sea level doesn't just require planning for impact and adaptation, but for when they will become uninhabitable, and eventually cease to exist. Among the first to go will be the Maldives,

Tuvalu, Kiribati (pronounced "Kiribas"), and the Carteret Islands.

The Maldives, a chain of islands in the Indian Ocean, is a major tourism destination, with a population of almost 400,000. It is distinguished as the country with the lowest elevation above sea level, less than five feet (1.5 meters). With most of the land area at three foot (0.9 m) elevation, this tiny nation could effectively disappear by the middle of this century. In the 1990's, a protection barrier was built, costing 63 million dollars. It did not work. The government is now in talks to relocate its population to either Sri Lanka, Australia, or India.

The 100,000 residents of Kiribati are also looking at a bleak future. In September, 2011, President Anote Tong gave a speech at the Pacific Islands Forum in which he mentioned the possible radical action of building structures resembling oil rigs for people to live on, at an estimated cost of two billion dollars. While the cost per person is not beyond imagination, what quality of life would that be? Would living on an artificial platform be less destructive to their way of life and their culture than transplanting themselves to higher ground, elsewhere?

UNPRECEDENTED LEGAL QUESTIONS

As nations continue to disappear there will be unprecedented legal questions. Although I do not claim legal expertise, it is easy to conjecture three rather novel categories of legal issues arising:

First, who has been hurt by rising sea level? While it will be obvious in the extreme when property disappears, there will be many more vague forms of damage. Whether a person suffers damage from sea level rise largely determines whether they can bring legal action. Related to this is the question of who is responsible for the damage. In much of the common law world, a breach of civil duty is called a tort, i.e. something done improperly or wrong. The most prominent tort liability is negligence. If the injured party can prove that the person believed to have caused the injury acted negligently, that is, without taking reasonable care to avoid injuring others, tort law can award compensation.

A potential precedent example of this kind of action has recently surfaced in the news. On September 22, 2011, the island nation of

Palau announced that it was seeking a hearing by the International Court of Justice, also known as the World Court, with the charge that greenhouse gas emissions are endangering its survival. Their application is in process.[117]

Second, there is a set of issues regarding where the displaced will be relocated and under what legal regime. In fact, there is already a term to identify them: Individuals Impacted and Displaced by Climate Change (IIDCCs). Many suggest that those displaced should be considered "climate refugees." Some experts contend that existing laws regarding traditional refugees are not adequate for this group of people.[118] Indeed, in September 2012, this concept was put to the test. As reported in a *Huffington Post* column, a man from Kiribati who had been living in New Zealand for years was denied refugee status there, despite the fact that he was unable to return to his native island due to sea level rise.[119] New Zealand authorities deemed that there was no such category under the Refugee Convention. I would imagine the policies will be revisited as the problems worsen.

The third category relates to the sovereign rights of a nation when it physically disappears. For instance, if the people of the Maldives are relocated to Australia, whose laws and judicial system must they follow, and for how many generations? If they are relocated, how is the political leadership maintained and defined over time? For example, if there is a subset of Maldivians living in Australia, do they continue to hold their own elections? In the case of monarchies, do the leadership rights continue as usual? For how long? How does one define emigration and immigration in such cases?

Each nation that votes in the U.N. General Assembly can affect the outcome of extremely important votes. Do those votes disappear when the land goes under, or could the people retain some virtual sovereignty even if they are relocated? Would living on artificial platforms entitle them to sovereignty? How large a platform is required? Must it be inhabited or defended?

What about the revenues that could continue to accrue to the sovereign nation as long as it exists? Two specific situations demonstrate the not-so-obvious importance:

a) The lease of fishing rights is both lucrative and important in terms of controlling policies for sustainable fisheries, which is more important now than ever. Do territorial waters, known as Exclusive Economic Zones (EEZs), continue to exist for displaced nations? Who qualifies to use that territory? Who gets to sell the valuable permit rights, and collect the revenue?

b) Tuvalu presents a special example of sovereignty rights. In the world of the Internet, each country was assigned one or more top-level domains. For example, in Britain, web addresses end in ".uk"; in Australia it is ".au." Tuvalu was designated ".tv." Soon thereafter, savvy marketers realized the value of website names with that suffix for internet television. Tuvalu began to reap millions of dollars a year from domain registration fees. Even though they sold out the rights in 2000, they continue to receive millions of dollars a year from the agreement.[120] Beyond that, there is a Tuvalu Trust Fund, supported by the international community, now worth roughly 100,000,000 dollars.[121] What happens to those financial assets if the country ceases to exist?

Beyond the geographic and legal aspects covered in this chapter, there are profound issues of culture, social justice, and the balance of power in the world. All of these are outside the scope of this work, but warrant serious consideration.

If you are elderly, wealthy, and do not have children, perhaps these practical and conceptual impacts are irrelevant. For most of us, the future of coastal property has daunting implications for our lifestyle, our economies, our feelings of security, and our moral responsibility not to cause injury to others.

The phrase "sea change" (first used by Shakespeare in Ariel's Song, in *The Tempest*) can describe any major transformation or alteration. It seems apt for the planetary metamorphosis that has just begun.

Chapter Thirteen
Property Values Go Underwater

National Public Radio did a series in 2010 about how the dramatic reduction in real estate prices from 2007 to 2009 had some beneficial effects, one of them being that many middle-class people could once again afford to buy homes.

As an example, they interviewed a Florida immigrant who had just bought a small house on Key West, where she worked as a maid. The lady featured in the radio story explained that her economic challenges were tough and that she had to work very hard just to survive. She was very glad finally to invest in America, and said that one day the house would be there for her children and grandchildren. When I heard her say that, I distinctly recall clenching my teeth, knowing that the outlook for sea level rise made her dream highly unlikely. Key West is situated on top of porous limestone, making it particularly vulnerable to sea level rise, as will be explained in Chapter 14. Furthermore, it is so low-lying that some of the city's storm drains already occasionally work in reverse, bringing seawater onto the streets during extreme high tides.

The house may very well survive physically for her children, but it is doubtful that it will be worth much over time. She is not alone. Millions of others in diverse locations will share her misfortune.

Property values will likely go underwater long before the property does. How soon will sea level rise impact property values? That is the crucial question. There are many factors that come into play in answering this. Let's take a closer look at them.

THE MARKET

Modern capitalistic societies generally view real estate as the safest place to store savings and wealth. This has been true throughout recorded human history. Even the law treats real estate differently than all other assets. With a steadily increasing world population and the basic rules of supply and demand, we were essentially assured that property values would increase every decade. And coastal land has been among the most valuable land in the world.

The time is soon coming when we will need to adjust our perceptions of real estate value to take the risk of sea level rise into account. That will be a fundamental shift in the way we value land, with huge repercussions.

Let's look at the problem, starting with the basics. Imagine that a decade from now you decide you want to live on the ocean. You find the right region and the right property. The future impact of sea level rise is widely recognized.

The current owners have had the property for generations. In the late twentieth century, they thought the acre on the coast was worth, say, one million dollars. But the early signs of coastal erosion are noticeable, even though it might be several decades before your desired lot is condemned. Several houses a few miles down the coast have already been abandoned. The property profile does not allow practical seawall protection.

In this situation, even if you had the money and really wanted that property, you would undoubtedly use good business and common sense to negotiate a lower price, knowing that the long-term outlook for coastal property was not good—might I say, *sinking*. The point is that markets adjust for known risks. We all do that, instinctively.

It is difficult to know when the discounting of property will begin. South Florida provides a good example because of its broad property market and high vulnerability to sea level rise. During the last decade, real estate prices increased wildly, as much as 25 percent per year, interrupted briefly by concerns during the heavy hurricane years of 2004 and 2005, and then led the national real estate crash in 2007.

Currently, there is still no sign that anyone in Florida is discounting the price of coastal real estate based on the slowly growing awareness that the shoreline will move significantly inland. At least for now, coastal property values continue to move with the larger real estate market.

Particularly in the era after Hurricane Sandy, there is growing awareness of coastal vulnerability. With increasing frequency people privately ask me, "If I sell within a decade, will I be okay?" It appears the concern is already developing, although no one wants anyone else to know. "Sandy" may have been a turning point in public awareness about coastal risks. Even though it was mostly a storm event, rather than a sea level event, there has been a noticeable change in public concern about this risk since that disaster.

It is just a matter of time for the inevitable vulnerability to be reflected in property valuations. Sea level rise data is already publicly available from several government authorities, in spite of the disastrous consequences the information implies for community tax bases, costs of adaptation, and in some cases, for the communities' very existence.

One might assume that the drop in property values will be sudden, once the long-term inevitability and magnitude are widely understood. I have now come to see that the drop in coastal values may be more gradual, perhaps as low as a few percent a year. Like guessing the stock market's response to an event, it is hard to know for certain how the crowd will respond. Two insights support the possibility of a gradual decline in values:

First, those who can afford the most expensive and appealing coastal lands may be willing to invest in them even if there is a risk of them disappearing in the coming decades. The personal appeal and value of being on the coast may simply be worth the risk and increased cost to them.

Furthermore, there is evidence that people rationalize real estate investments over roughly a 30 to 50-year period, mostly based on subconscious, emotional reasons. In fact, some economists have justified this, citing that even at a discount rate of a few percent a year, a property is devalued to negligible levels within about 50 years.

THE INSURANCE FACTOR

One would expect the cost of coastal flood insurance to increase due to the risk of rising sea level, but the system is not so straightforward. In our society, insurance rates are subject to regulation, and perhaps also to politics. The cost of insurance in a vulnerable zone may actually be split into different policies that are hard to distinguish. In coastal areas, basic insurance may cover fire and liability. Windstorm insurance may be a separate policy to cover damage from hurricanes and similar causes. In the United States roughly seventy percent of the flood insurance is covered by the National Flood Insurance Program (NFIP), which is part of FEMA (Federal Emergency Management Agency).

Once again, Florida gives us a preview for how changes in insurance can play out in a political environment. Consider the following sequence of events:

- Florida has a long history of vast hurricane-caused coastal damage. After insurers faced a series of major losses in 2004 and 2005, they tried to raise rates.

- The state denied the rate increases requested by the insurance industry.

- In 2006, a large segment of the insurance industry essentially pulled out of insuring the high-risk coastal property areas of Florida.

- The Florida state legislature responded by establishing a government-sponsored company to take over, Citizens Property Insurance.

- As many suspected, the state was subsidizing rates. In the summer of 2012, Florida announced that it would start

> forcing a majority of the state-subsidized policy holders back into the private market to reduce the state's risk exposure. Initially, there was a lot of confusion, complaint, and as was inevitable, higher insurance rates. It is too early in the process to know how this will unfold. Legal actions by policy holders have just started.

We should also recognize that there have been voices of enlightened concern about the risk of sea level rise to Florida. In 2009, a clear and bold report was produced, "Florida's Resilient Coasts," with a large cover endorsement by former Governor Charlie Crist. It pulled no punches and pointed to the solid science behind rising sea levels, the inevitable damage, and the need for decisive policy changes. It was a step in the right direction, but went largely unnoticed. His successor, Governor Rick Scott, has taken a very different stance, partly out of financial necessity, but also apparently out of the low priority he places on issues related to climate change.

Policy tends to be made according to very short-term interests. Big change seems to come only when we have no other options. With something as slow and steady as sea level rise, getting politicians to act will be a challenge, though again, "Sandy" seems to have initiated some change in attitude.

It does seem inevitable that eventually insurance costs will significantly increase in all coastal areas to cover the inescapable risk. The National Flood Insurance Program (NFIP), mentioned above, has been tapped heavily since its inception in 1968. This program extends from barrier island coastal communities in Texas all the way around Florida to the Carolinas and farther up the coast. Inland, coverage extends well beyond the huge floodplains along the Mississippi to North Dakota. In June, 2012, over 5.5 million Americans carried an NFIP policy, with a total insured value of 1.2 trillion dollars.[122]

The program was originally meant to be self-funding. The annual premiums were supposed to cover the administration and average annual loss from payouts. Its purpose was to encourage communities to reduce risk by enacting better standards for building codes, elevation requirements, drainage systems, and zoning. In many instances, this has happened. However, like most government programs, over the years it has grown in scope, benefit, and overall cost.

The huge increase in flooding disasters in the last few years appears to have prompted FEMA to privatize the NFIP. An industry insider told me that five different major insurance firms met to discuss taking over the program. As they evaluated the portfolio, it became clear that the flood insurance policies were badly underfunded, to the tune of about 19 billion dollars at the start of 2012. The negotiations stalled when the government made it clear that they did not intend to compensate the private companies for taking over the program.

That essentially ended the negotiations. Just as with the Florida "solution" to high insurance costs, the NFIP dilemma seems to underscore the danger of politicians trying to get involved in the risk market, however appealing it may be to find a political solution.

NFIP's parent agency, FEMA, typically requires emergency replenishment by Congress after major disasters. The problem is a direct result of dealing with an increasing number of natural disasters. When faced with these immediate crises, it is hard to step back and ask if there is something fundamentally wrong with the program. Compassion and politics prevail in the short term, but at some point, the financial aspects will force us to confront the issue.

The US Congress did pass some reforms to the NFIP in 2012 to begin to limit the payouts, subsidies, and public exposure. Those changes were just beginning to kick in on October 1, 2013 as this second edition goes to press. Various provisions are dramatically increasing coastal insurance rates in certain situations, in a few rare cases actually boosting premiums by a factor of ten. In just a matter of a week or two there are howls of complaint about the economic hardship. Already elected representatives are calling for relief.

This will be an evolving saga of economics, politics, and geologic destiny. In the short term, it is anyone's guess how it will play out. In the long term, there is little doubt that policies and economics will have to reflect the growing risk.

In my view, there are three types of risk that we should differentiate. First, there are truly unpredictable catastrophic events like tornados that destroy entire towns. Second would be probability events like hurricanes, which have risks that can be calculated similar to the risk of a fire or car accident. Finally, I believe that the long-term risk of rising sea level and coastal erosion must be a third category, because it will definitely hit most coastal communities, and because the demands for compensation and rebuilding will be so widespread that no government will be able to foot the bill for any sustained period.

A 2009 study in California estimated that the anticipated 4.5 foot (1.4 m) sea level rise this century would destroy 26,000 acres (41 square miles) of coastal land by erosion just in that state.[123] Using a crude estimate of two million dollars an acre, that translates to something like 50 billion dollars worth of the land, not counting the cost to replace the structures and infrastructure. Similar situations exist all along the U.S. coastline, and internationally.

How far can we expect the national or international government agencies to go in terms of underwriting these risks? As I write this, the U.S. national debt is at some 14 trillion dollars; numerous other countries face huge debt crises and even insolvency. There is a real question of how much risk and loss governments can absorb, in addition to the burdens of national security, social and medical programs, infrastructure, administration, and debt service.

Our culture has developed the expectation that government will somehow take care of us when natural disasters occur. Regardless of your political leaning or philosophy about the role of government, I submit that our budgetary challenges, combined with unprecedented long-term sea level rise, may overturn a lot of what we have come to expect. At some point federal and state governments will undoubtedly have to rely on the risk assessment of a competitive insurance industry.

The key question for anyone with coastal interests is whether you can count on the federal government to cover your assets whenever the disaster does hit. Inevitably, governments will not be able to compensate everyone for the risk of assets exposed to rising sea level. The key questions are when that will happen, how much notice you will have of a change in policy, and whether you will then be able to liquidate your assets at a reasonable price.

ACCOUNTING AND BANKING IMPLICATIONS

When vast sums of money are committed to buildings and infrastructure, banks assign those structures a limited useful life, and amortize them accordingly. Often these major built assets have a much longer life and are still useful a century or more beyond what was originally estimated. Iconic examples in New York City, for instance, would be the Empire State Building, Rockefeller Center, and the Brooklyn Bridge. There is huge economic and practical benefit from this extended period of residual utility.

We are now entering a new era, where sea level rise will make the functional life of structures much more finite. When we have exhausted the effectiveness of seawalls and modest defensive measures against sea level rise, coastal buildings and infrastructure will be abandoned. That is not an "if" but rather "when."

The cost of this eventual abandonment will be absolutely enormous and will include:

- Accelerated amortization/write-off of structures and infrastructure.
- Write-off of the underlying real estate, something that is rarely done on balance sheets now.
- Significantly increased insurance costs.
- A dwindling tax base.
- As things deteriorate, the increasing cost of defensive measures such as seawalls, and then the costs of remediation and repair as things get destroyed, prior to the decision to abandon.
- The cost of relocating, not just individual buildings and infrastructure, but entire communities.

Amortization (the reduction in value) of coastal assets will be very challenging. It will require implementation of important new concepts from the accounting profession that will affect financing, taxation, and insurance. The formula and/or assessment of property, particularly for major assets, will be complicated, likely taking into account anticipated global sea level rise over the lifespan, land elevation, subsidence, the particular structure, and even the outlook for the community since that too affects market value. Such a fundamental change in generally accepted accounting principles will affect everything from public companies, to personal investments, to estate planning.

Having talked to experts in accounting and economics since this book was first published, it is clear that this will be a challenging change for their professions. By present rules, the slow gradual write-off of assets does not meet their tests of sufficient impact within five years. Sooner or later the professions will awake to the new reality.

Financial institutions and real estate investors could take huge hits from the amortization of coastal assets. It is questionable whether coastal properties will be viable for the fairly typical 30-year mortgage terms. Institutions or individual investors holding "long-term paper" secured by coastal assets will want to look at the potential risks of such investments.

Even more vulnerable are the property owners, since their equity usually will be wiped out before that of the mortgage holder.

It is worth pointing out that vulnerability is not limited to those right on the waterfront or at sea level. If rising sea level undermines the value of shorefront buildings, inland buildings can easily be affected. In fact whole sections of communities can be affected.

Rarely will cities be abandoned suddenly. The withdrawal will happen over many years and will likely be chaotic and painful. Like in an old, decaying neighborhood, there will be early leavers, those that hang on until things get grim, and then those that defiantly say they will never leave. The difference with the encroaching sea is that there is almost no possibility of turnaround and rebirth within human timescale. As explained, the heat dynamics of the warming atmosphere, the heat stored in the oceans, and the melting of the ice sheets strongly indicate that sea level will continue to rise and will not go down for at least a millennium.

"Buy land, they're not making any more of it" has been sage investing advice for generations. The point of that wise guidance was that land was the only thing you could count on with certainty. The new reality that rising seas will destroy a considerable amount of our most valuable real estate may warrant a new slogan.

What We Can Do

"It's far too late and things are far too bad for pessimism."

~ Dee Hock

"In every deliberation, we must consider the impact on the seventh generation..."

~ The Great Binding Law of the Iroquois Native Americans

Chapter Fourteen
Defend or Retreat

Military strategy looks at the question of when to advance, hold a position, or retreat to a more defensible location. Long-term rising sea level presents a somewhat different adversary. The growing realization that the sea will rise for centuries, punctuated by episodes of severe storm damage, presents a new question about holding the sea at bay or finding safe ground. Anyone who has witnessed the power of the pounding surf in a stormy sea will appreciate the immense force at work.

The options likely will not be clear or easy. Moving could mean moving back from the sea or abandoning the area entirely. Another possibility is to move up, like Galveston, Texas, did.

GALVESTON'S 17-FOOT SEAWALL

The Galveston Hurricane of 1900 was this country's deadliest, killing 6,000-8,000 people. Recognizing its vulnerability, the city subsequently settled on a bold plan to build a massive 17-foot-high seawall that now spans some 10 miles (16 kilometers).

Engineers also essentially lifted the city to a higher elevation. In addition to the seawall, some 2,000 buildings were raised a similar amount. This smart response was most likely a good investment for two reasons: it gave the city a fairly secure future for more than a century, and the seawall itself became an attraction, bringing considerable revenue to the community.

Even with the current sea level rise forecasts, a great seawall such as Galveston's could extend a city's viability by a century or more. Still there are limitations to this approach.

Spending the hundreds of millions of dollars it would cost for such a project today has to be evaluated in light of its impact on quality of life and existing infrastructure. How will it affect local businesses and homeowners? Will it interfere with access to fresh water supply, road access, railroads, and utilities? Where does the wall end? What happens if the sea gets around or behind the wall?

Extending the life of a single community becomes a different proposition if the adjacent communities abandon parts of the shore to sea level and erosion. Eventually, the protected community will become isolated and the cost of daily life will increase as the number of residents slowly declines.

BARRICADES—YES OR NO?

Defenses against the encroaching ocean can take many forms, including seawalls, bulkheads, revetments, retaining walls, dikes, and levees. They can be vertical, curved, or sloped, made of concrete, rock, steel, or earth.

Over the past century or so, we have mostly built these structures to prevent erosion associated with the natural movement of beaches, to utilize low-lying land, and to prevent a recurrence of storm damage. In recent years, however, the realities of rising sea level are causing communities to look at defensive barriers the way passengers on a cruise ship might pay attention to its lifeboats.

If the goal is to extend the useability of an area for a couple of decades, it may be realistic to invest in an ocean barrier of some type. But such structures are expensive and have real limitations when it comes to the range of sea level rise, combined with storm surge on the higher shoreline. Typical storms are bad enough. Rogue storms—so-called hundred-year storms—are now happening more frequently, often more than once a decade. In fact, we are increasingly experiencing what once might have been viewed as thousand-year storms.

Weather patterns are changing, and communities must consider longer timeframes—several decades or more—in determining what if any barriers will make their homes, businesses and communities livable over the long term. Here it is important to recall the distinction that storms and sea level rise may require different defenses. For example a good line of sand dunes is a barrier for storms, but not rising sea level.

When faced with spending hundreds of millions of dollars or more on barrier systems, how much sea level rise do we design for? Are we looking at a three foot rise this century? Six feet? Ten feet? Twenty feet? The cost difference in the designs is huge. If the foundation is designed for three feet, and eventually you require 10, the foundation will not be adequate.

The legal issues of seawall construction must also be considered. If a governing authority creates, or even just approves construction of a barrier, it can be held liable if that structure fails. The inevitability that many barriers will fail over the long term, creating serious liabilities, could even deter local governments from approving projects. Communities will struggle as property owners and business owners fight to stay where they are and try to keep things viable for as long as possible.

Many cities, states, and nations are already facing fiscal insolvency. Consideration of sea level rise should be a serious red flag for future development, or even for maintaining current status. The dilemma becomes even greater as we look a few decades into the future.

Let's imagine it's 2050, just four decades away. It is perfectly reasonable to envision a foot or more of sea level rise, assuming nothing catastrophic has occurred in Antarctica. By that time, we'll also know with greater certainty whether the current projections of sea level rise for the end of the century are on track or whether we should expect a quicker pace of rise.

Recall that 14,000 years ago the sea level rose more than a foot a decade, continuously, for four centuries. The warming in our present era is hundreds, if not thousands of times greater. By the middle of this century there is a very real chance that we will be looking at a much more ominous forecast, possibly the catastrophic sea level rise associated with the thawing of the Western Antarctic ice sheet. How high a wall do you build when you know that the rising sea will most likely accelerate to the point of a foot a decade by the end of the next century?

If we accept that within a further 100 years—roughly the year 2150—we could be facing 20-30 feet (6-12 meters) of higher sea level, the prospect of retaining walls holds little appeal.

Our perspective here is not to forecast doom and gloom, but to allow individuals, investors, corporate leaders, and civic officials to have a context in which to plan. As we all know, it is easier to make short- and medium-term plans if you know with relative certainty where things are headed in the longer term.

Porous Limestone – The Achilles Heel

There are many reasons why building a high seawall may not be possible in many locations. One that has been referred to already in this book, but is barely recognized, is that many tropical areas are built on porous limestone, the remnants of ancient coral reefs. This type of foundation rock is common on islands in the Pacific, many in the Caribbean and Bahamas, and in places like South Florida.

Compared to granite and marble bedrock, which are essentially impervious to water, porous limestone is like a sponge with very small pores. That porosity makes a seawall around cities like Miami completely unfeasible. Aquifers and groundwater are based on the flow of water through such porous rock. Water can flow through the rock in immense quantities for hundreds of miles.

To appreciate this tragic vulnerability, let's imagine a limestone-based island, which is threatened by rising sea level. Even if you built an impermeable barrier around the island with a very deep foundation, the sea level would rise to the same level inside the wall as it did outside.

Parts of Miami and other areas of southern Florida are already struggling with the realities of this geologic base. During storms and extreme high tides, even low-lying areas far inland experience water coming up from the ground as spontaneous puddles. The water slowly dissipates when the ocean level recedes, even if the shoreline is miles away. There is no known technology by which communities built on porous limestone can survive higher sea level at their current elevation. Their only option is to lift their communities upward, literally, like Galveston did a century ago.

A good friend of mine who is a leader in the climate change arena recently told me a personal story that underscores Florida's dilemma. Daniel went home to visit his mother in Miami Beach, where she is a highly respected judge. He noticed pooling water in her front yard and asked her about it, since he saw no signs of heavy rain. She seemed unconcerned and said it only happened a few days a month, though some months were worse than others. He recalled that did not happen 25 years ago when he was growing up there. When his mother confirmed that to be true, he quickly realized the pooling water was an early sign of sea level rise, as pressure from the rising sea forced water up from underground to form puddles in her yard. Even a sharp judge did not realize the significance of what was happening right in front of her because it was happening so gradually.

It is hard for us to recognize slow creeping change. It's like the proverbial "boiling the frog" story. If a frog is placed in hot water, it will immediately jump out. However, when put in the pot while the water is cool and slowly warmed, the frog is fooled by the gradualism, eventually boiling to death.

Moving Back Or Moving On?

In 1962, Hurricane Hattie destroyed 75 percent of Belize City, the coastal capital of Belize, which was then known as British Honduras. Belize is a tiny country, just south of Mexico's Yucatan peninsula, and north of Honduras. What makes it especially vulnerable to hurricanes is that it is an "elbow" of the Western Caribbean. Even without a direct hit, hurricanes tend to pile up extreme sea levels in such pockets.

After the devastation of *Hattie*, the country successfully built an entirely new capital, Belmopan, on higher ground. Belmopan was built far from the coast at an elevation 250 feet (76 meters) above sea level.

This small Central American country had the common sense not to keep investing in harm's way. Though their primary risk concern was hurricanes, the location they chose is essentially above the ultimate high tide mark, even when all the ice on our planet has melted.

While it was undoubtedly easier to relocate the capital of Belize than say, Miami, Charleston, or New Orleans, the "Belmopan Solution" can be one model among other farsighted approaches that demonstrate a community's understanding of the long-term impact of sea level rise.

Retreat

In the many coastal communities where rebuilding is simply not an option, abandonment, often described as "retreat," will be the excruciating choice to be made.

Retreat goes against the nearly global cultural heritage that has shifted from a nomadic lifestyle to the settled existence we have enjoyed for thousands of years. It also suggests a defeatist attitude, repugnant to many.

Yet, as more people awaken to the future struggles facing their coastal communities, retreat will occur. It will be anything but logical and orderly.

A small percentage will quietly leave early, hoping to sell and relocate before real estate values drop. Eventually, coastal communities will begin to organize among themselves to prepare for the coming crisis. Interim solutions will likely include new building codes, new coastal setbacks, and zoning changes. There will be plans to accommodate infrastructure, particularly water supplies, wastewater treatment facilities, power, and essential transportation routes.

People have so much invested in their homes and their communities that these will be wrenching discussions. Just as some will leave in the early years, others will hold on as long as possible.

Those who do stay will find increasing costs. At some point the communities will start to slowly get smaller, both in terms of useable space and in terms of population. The scenario will take place in many acts, perhaps over 50 years or more. This is essentially what happened on Sharps Island and Holland Island, the two places in Chesapeake Bay that we covered in Chapter Six.

Complete abandonment will be the last step, many decades from now. One can well imagine local leaders warning that the community will soon cease to provide essential services of water, sewage treatment, police, and fire. A small percentage of the homeowners, particularly those who developed elevated properties and who own boats, may just stay anyway.

If you think this is fantasy, read about the Stiltsville community in the waters off Miami. On Wikipedia and elsewhere there are fascinating pictures of this group of houses built on stilts. Some survive in spite of decades of government efforts to eliminate the structures.

So far we have been largely portraying and visualizing the question of defend or retreat in the context of the developed world. There are also tens of millions already threatened by sea level rise in the developing world. In places like Kiribati in the Pacific, mentioned previously, there are already houses partially underwater. In low-lying Bangladesh there are now tens of thousands whose houses are awash during monthly lunar tides. As their daily lives are already consumed by the sheer struggle for existence, they will avoid the expense of abandonment as long as possible. Even in the United States, those in a daily struggle for survival may find it virtually impossible to afford the cost of relocating, particularly when they realize that their homes no longer have value.

For all of us, seawalls and other defenses will have their limitations. Eventually, everything along the coast will require adaptation of one form or another. Like most realities, the sooner we embrace the situation, the better we can deal with it.

Chapter Fifteen
Intelligent Adaptation

Albert Einstein said, "The measure of intelligence is the ability to change." Rising sea level will present us with boundless opportunities to demonstrate our intellect over the next half century. Among the many challenges will be separating practical solutions from the fantastical and unrealistic. To start, let's look at an example of the latter.

Floating House Nonsense

Floating houses have drawn recent media attention in places as diverse as Amsterdam, London, and Seattle. The futility of this approach is perhaps most obvious when we consider the associated costs. A durable houseboat for saltwater use is considerably more expensive to build than a basic land-based house. Greater Miami has 2.5 million people. Greater London, more than eight million. In Bangladesh an estimated 20 million people are in jeopardy. Building floating houses at those scales would be very expensive.

Even if you could build floating houses for these populations, the cost per user to maintain their essential services is significantly higher than that of land-based conventional houses. As any boater knows, saltwater environments are extremely harsh and corrode nearly everything. Maintenance costs are high. Bringing utilities to a floating house would cost considerably more than bringing them to a typical residential street.

While it might be possible to maintain 500 or even 5,000 floating homes for the affluent or in a lagoon separate from the ocean, does that constitute a "solution" to sea level rise for the threatened cities? What about the sewage plant, garbage disposal, every shop, police station, and other aspects of infrastructure required for a community, both on routine days, and particularly during stormy weather? The cost of these services would only increase as the size of the community shrinks over time.

Finally, there is the vulnerability to storms. If you consider the mighty force of crashing waves and the surges from increasing numbers of storms, it becomes very clear that floating houses are not the solution to rising sea level, given the scale of what lies ahead.

Real adaptation to sea level rise must be *intelligent*. Anything else is silly and distracts us from the thoughtful solutions that individuals, businesses, and communities need to consider.

INTELLIGENT ADAPTATION

Intelligent adaptation to sea level rise will require us to shift our perspectives and expectations, and to have the courage to make difficult, long-term choices. I suggest that the process should consider these five points:

1. **Act with a long-term perspective.** We should not naively build protections based on short-term increases in ocean height. Instead, we must consider the long-term picture of where things are headed, with a responsible look at the "return on investment" for spending money on property, infrastructure, or improvements. No amount of wishing it were otherwise will change the fact that *ultimately* our low-lying coastal cities will need to be abandoned. Whether that is decades or centuries from now, this reality needs to become part of our new awareness.

2. **Accept that there is a range of projections this century.** We should act on the best scientific consensus about sea level rise, as it

exists today, knowing that forecasts will shift as each decade validates or causes modifications to our projections. It's interesting that many people, particularly those who make business and financial decisions, are reluctant to act on less-than-perfect information about climate change and sea level rise. Yet those very same people make decisions all the time about investments in companies, stock markets, products, commodities, interest rates, and currency fluctuations based on a range of expectations, and a broad range at that. We must use the same standards when thinking about the range of projections for sea level rise over the course of this century.

3. **Consider geology as well as topography.** The right approach to adaptation in Miami is entirely different from that for Manhattan. While a challenge, Manhattan's geology makes it possible to plan for a century or more. South Florida and many islands with very low elevation and porous limestone face a much more devastating impact as the decades progress. Every coastal location needs to consider its specific topographic and geologic factors, which may include land subsidence, tidal amplification, and storm surge vulnerability.

4. **Recognize the finite future of government bailouts.** In the US and in much of the world, we assume that governments will subsidize, bail out, and reimburse us for almost any natural disaster. It is certainly compassionate and desirable when looked at from the perspective of those affected. But the system will simply break if we do not have the wisdom to change our policies. Sooner or later we will have to treat unexpected natural disasters differently than property loss from predictable, permanent, universal sea level rise. It is economically inevitable.

5. **Anticipate property devaluation.** It is almost certain that coastal property values will go "underwater" long before the water actually hits the land. The road downward will likely be uneven, sometimes just a few percent a year, perhaps sometimes 20 percent or more in a single year. Fluctuations will be based as much on emotional shifts as on tangible ones. The timeframe for decline might vary from a few years in some very low-lying small coral island, to a

century in places where a solid geologic structure under a major city might support sophisticated adaptation measures. Forward-looking companies and individuals will begin to provide for this similar to the way a building is written off gradually over time as part of routine accounting and finance practice.

THINKING BIG

Intelligent adaptation covers not only a decision of what to protect, but also recognition of the magnitude of change and timeframe for action. In the short term, there is no need to over-react. In the long term, it would be hard to overreact. We will need to apply all of our human ingenuity toward bold engineering solutions.

Imagine something on the scale of controlling the water level of San Francisco Bay, the Chesapeake Bay, Mediterranean Sea or The Baltic Sea. As outlandish as it may sound, huge engineering constructs might be applied to any of these four large, semi-contained areas.

Needless to say it would be a daunting challenge to close the Mediterranean at the Straits of Gibraltar given its span of more than eight miles (14 km). A Baltic Barrier would be even more challenging with the span of some 40 miles (60 km) at the Kattegat between Denmark and Sweden. The Chesapeake already has the 17-mile (28 kilometer) bridge/tunnel system across its mouth that might be a basis for a full barrier. The Golden Gate bridge at the entrance to San Francisco Bay is perhaps easier at 4,200 feet (1,280 meters) across and such a barrier is now being discussed publicly in the Bay area.

Any of these would be enormous engineering feats beyond anything done to date. Yet I think they will eventually be considered once the reality of rising sea level becomes unavoidable. There must be dozens of other locations globally suitable for bold concepts that would create a new stable environment in spite of much higher sea levels.

Thinking about solutions at this scale lifts our vision above the usual concepts of small, localized seawalls. In these specific locations, such boldness might preserve near-present sea level for at least a century, perhaps much longer.

Blocking any of these bodies of water to maintain sea level will also require dealing with the inflow from rivers. That 'problem' could also present an opportunity to divert some of the in-flowing fresh water, that will be sorely needed for agriculture and human consumption.

The Dutch Delta Works was intended to reduce the risk of flooding to once in 4,000 years. when it was designed a half century ago. The largest segment is the Oosterscheldekering, shown above, spanning 6 miles (9 km). It is one example of how a country is protecting a large bay from the rising sea and tides. By 2008, higher sea level projections caused then to initiate a massive enhancement program under a new Delta Works program. (Photo, courtesy of Source: http://www.deltaworks.org/Author: Delta Works Online - Job van de Sande.)

Throughout history mankind has come up with remarkable feats of civil engineering to solve perplexing problems. Looking back to ancient times we might cite the Roman aqueducts, or the tremor-proof building techniques of Machu Picchu. In the modern era there are thousands of examples in aviation, space, and road infrastructure. Just a few centuries ago, cholera plagued the cities until the concept of modern sewage treatment and sanitation was implemented. More relevant to the challenges of sea level rise might be the huge dams from Aswan to Three Gorges, the flood control levees along the Mississippi River, the Thames River Barrage in London, and the ocean defenses of the Dutch Delta Works project.

The Dutch Delta Works is considered to be one of the seven wonders of the modern world by the American Society of Civil Engineers. It is a vast array of dams, sluices, levees, locks, dikes, and storm surge barriers (see above image). For a half century, the Delta Works system has allowed the Netherlands to protect itself from occasional storm surge and the small increase to date from sea level rise. About a third of the country is below sea level making coastal barriers essential.

Recognizing the new reality of accelerated sea level rise, the Netherlands is now examining enhancement options, since the Delta Works will not come close to maintaining the original design criteria of protection from a once-in-4,000-year weather event. Parts of the country may not be able to be protected at all within the next century or so.

Throughout the world, there are locations that are not geologically suited to protection or may not have the economic value and population density to justify the tens of millions to billions of dollars that would be required for effective protections. If the melt rate follows the non-linear curve that is predicted by Dr. Hansen and others, there will eventually come a time when the rate of sea level rise overcomes many of the solutions.

Whether that happens in 50 years, 100 years, or 150 years will make a difference in our approach. Based on the information we already have, we can begin the evaluation process now. We must start to look at the situation much more holistically than we did a century ago, when we started burning coal without much consideration for its consequences. Things are changing quickly. The sooner we begin to apply our inventiveness and technology to adaptation the better.

PIONEERS

Some communities have started taking intelligent steps towards adaptation. Three very diverse examples are:

- The San Francisco Bay Conservation & Development Commission has done an extensive study of the options to accommodate a projected sea level rise of up to 55 inches (1.4 meters) this century, "San Francisco Bay: Preparing for the Next Level."[124] Arcadis, a Dutch-American firm specializing in this field, was the lead consultant in the effort. The result was an outline of four alternative paths for future development and adaptation, characterized by the simple comparative images shown in Figure 15-1.

San Francisco Bay Development Scenarios

1. Tidal Denying Development

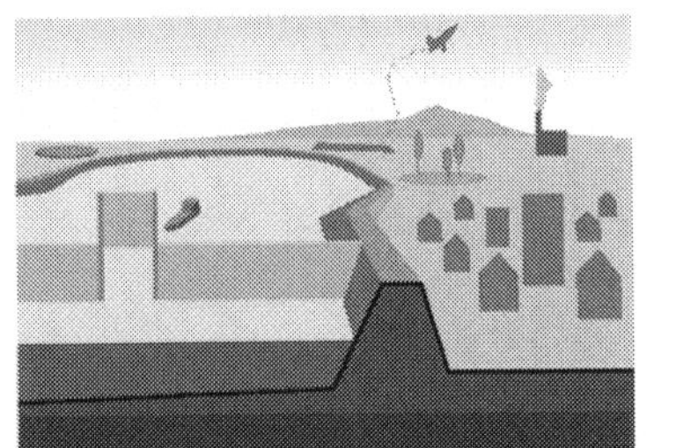

2. Economic Driven Development

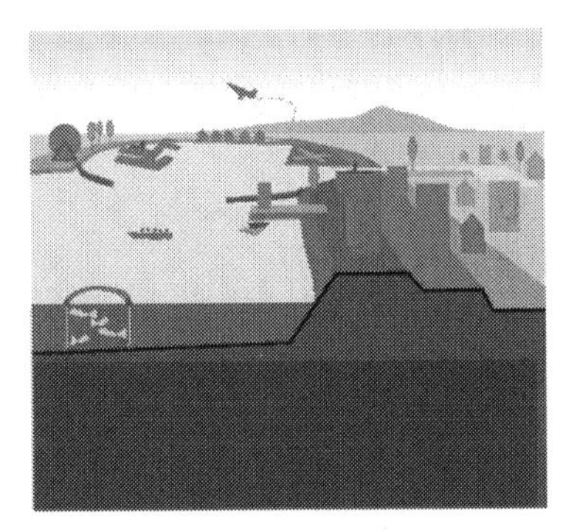

3. Tidal Embracing Development

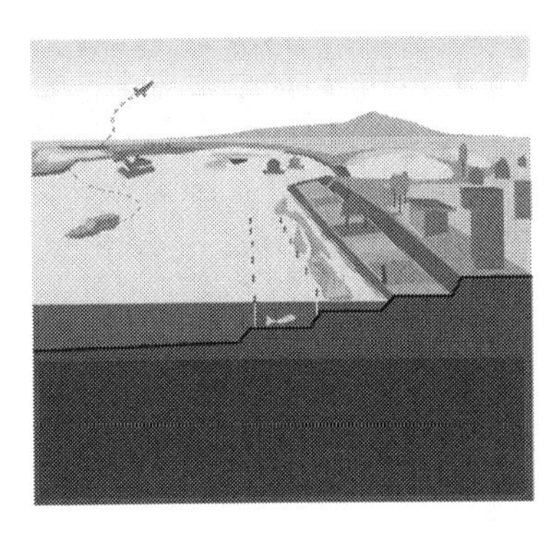

4. Ecology Driven Development

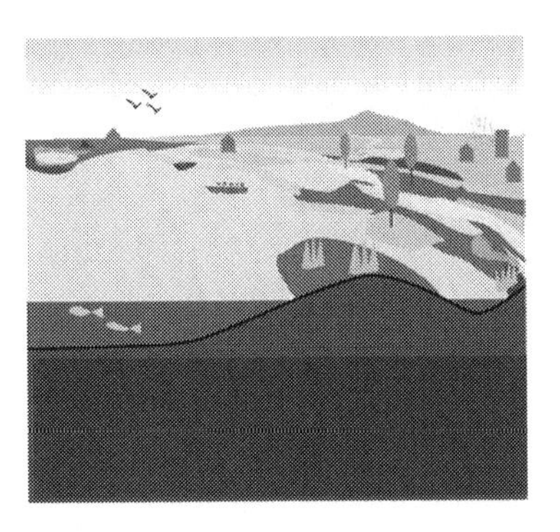

Figure 15-1. San Francisco Bay is one area in the United States that has diligently looked at long-term scenarios,taking into account inevitable sea level rise. These four illustrations characterize the different approaches to dealing with sea level rise beyond four feet. More information is available at: www.bcdc.ca.gov and www.adaptingtorisingtides.org/. (Courtesy of San Francisco Bay Conservation and Development Commission.)

- In Australia, the state of New South Wales has taken a clear position that sea level rise of almost a meter should be expected this century and that all future building permits and zoning should recognize that. They went further by requiring properties located in the projected flood zone to identify themselves as such when listed for sale and also on the title documents themselves. In 2011, this forward-looking move elicited backlash from angry property owners who chafed at the impact on their property values. The battle lines are just being drawn. Time will tell if the state government is ready to act in the long-term or more immediate interests of its residents. The same battle is starting to be played out in a number of other locations.

- Boston held a national competition to invite visionary architectural ideas for the city, looking ahead a full century. In one of the categories, first place was awarded to Antonio DiMambro and his team at Comunitas (now renamed **DiMambro** + Associates). Their bold concept, titled "Boston's Safety Belt," would adapt to rising sea level by connecting the harbor's barrier islands. Clever "gates" would allow shipping and recreational boating, but would keep rising sea level outside of the new harbor lagoon. The plan would create many related benefits, including large areas of new valuable waterfront real estate. From what I have seen, this is the best example of a creative plan that could ultimately create value rather than added cost for a community. What is amazing and sad is that the contest and award happened in 1988, more than two decades ago, but the plan gathers dust due to a lack of funding and political prioritization. (See the 2010 *Boston Globe* article about this at johnenglander.net/boston.) The plan and cross section are shown in Figure 15-2.

Several firms in the U.S. and internationally are working with other communities to figure out what type of engineering solutions could work. In the first edition of this book I named a few firms that are already working on the challenge. Since that edition a number of other firms have contacted me and impressed me with their expertise and resources. This will be a changing field, so I will refrain from naming such firms in this book format. (Interested companies or communities are welcome to contact the author directly.)

Boston's Safety Belt

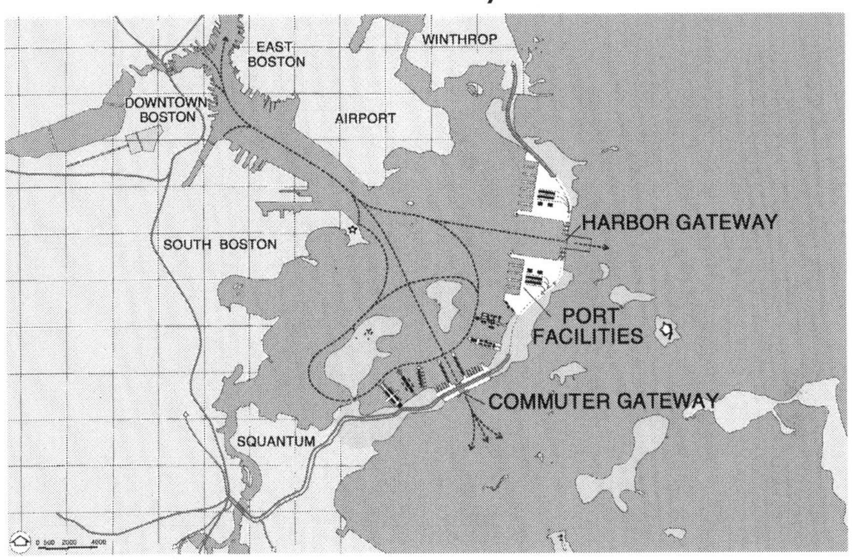

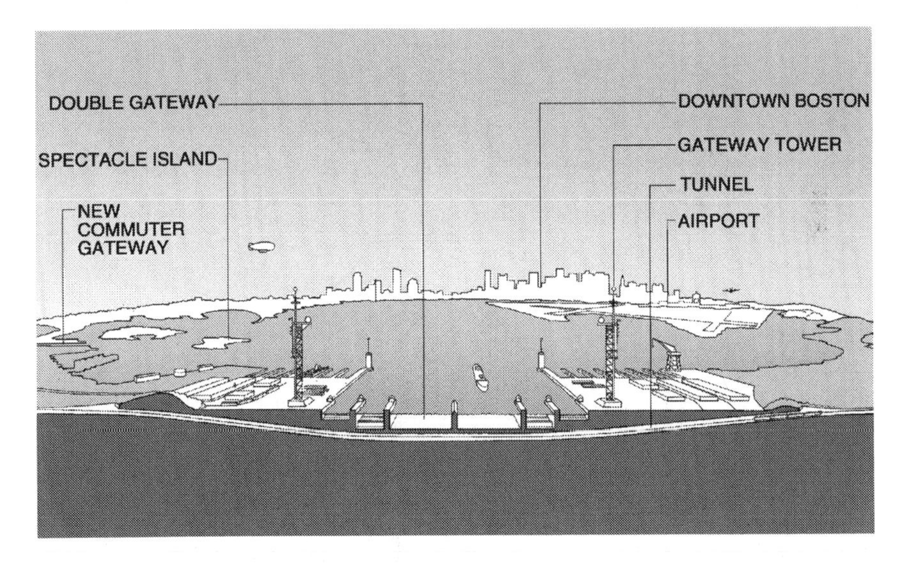

Figure 15-2. The plan and cross section above are from "Boston's Safety Belt," an innovative solution to the problem of rising sea level in that highly vulnerable city. This forward-thinking concept could result in a very positive return on investment and good quality of life in the face of considerable sea level rise over the coming century. In spite of winning first place in a design competition in 1988, the plan gathers dust on a shelf. (Illustrations courtesy **DiMambro** *+ Associates.)*

Further action will require more public support. Perhaps books like this, combined with articles in the press and public discussion, will push more communities to be proactive.

THE SILVER LINING

While sea level rise will present truly massive disruption, there is at least one bright side to the situation. Most of the time, we don't have the opportunity to anticipate catastrophes. Deadly hurricanes give us a few days to prepare; tsunamis may allow hours if warning systems work; tornados give us minutes; earthquakes can give no time at all, aside from the vague general forecasts along the lines of "seismologists foresee the significant possibility of a major quake some time this century."

With rising sea level, however, we have decades, even a century to plan, prepare and adapt. Present technology allows us to accurately predict the areas at risk of flooding and storm surge decades in advance. We do not need to panic, but we do need to act.

There is another aspect of sea level rise that can be seen through a positive lens. I am reminded of an observation made by my long-time friend and one of the most widely recognized marine experts in the world, Dr. Sylvia Earle. She notes that the current generation is the first to see the immense impact we are having on the life-giving sea. Satellite imagery and Internet connectivity allow people all over the world to understand how deeply humans have altered the health of the vast ocean and the web of life within.

She exhorts us to convert that awareness to the challenge of our age: to do what is necessary to preserve sustainable seas for our children, grandchildren and beyond. Sylvia's message about marine ecologic sustainability also applies to climate change and sea level rise. We are at a unique juncture in history, when we have the awareness of a problem and the intellect to correct things. It presents us with the large question of whether we will step up to the challenge. That also brings to mind the famous line, "If not us, who? If not now, when?"

Dr. Joseph Bouchard, the former commander of Naval Station Norfolk, is a good example of those who are now quite vocal about what our priorities should be. Using the potential damage to Virginia military bases as an example, Bouchard said:

> *It could eventually cost more to rebuild damaged infrastructure and protect against higher waters than it would to change energy-use patterns now.... It's important for what I would call very conservative reasons: protecting our economy and protecting our national security.*[125]

Some readers will recall the underwater explorer and pioneer, Jacques-Yves Cousteau. He vastly expanded my generation's perspective about this ocean planet. During the course of his nearly nine decades of life, he contributed to huge technologic advances, travelled to most countries on all seven continents, and had a towering influence on the twentieth century. Not only did he invent scuba diving equipment, but he shared his discoveries via *The Undersea World of Jacques Cousteau*, back when there were only a handful of television channels.

Just months before he died in 1997, he and I spent a few days together and had some lengthy and even late-night conversations. The topics ranged from the oceans to population, climate change, technology, quality of life, and our political world. There is little question that Jacques was a genius who saw the world through a different lens. My notes reflect four of his insights strike me as relevant to this issue:

1. We are on the verge of some environmental and climate disasters, which could demonstrate humans' capacity for greatness or lead to our demise.
2. We have an obligation to improve the quality of life for people throughout the world.
3. Mankind has little choice but to use technology to try to fix the problems we have created.
4. It is up to citizens and organizations... to advocate political solutions or at least to identify principles and goals. If we set the agenda and build support for it, the politicians will want to "join the parade."

That sage advice is worth considering as we ponder what each of us can or should do.

Chapter Sixteen
New Technology and Political Will

While we can no longer stop sea level from rising, it is urgent that we do everything possible to slow it, given the catastrophe that lies ahead if we stay on our present course. Mitigation is the *wonkish* scientific term for this, though I prefer simply to talk about *slowing sea level rise.*

To slow the pace of sea level rise, we must slow the temperature increase created by greenhouse gases. Tens of thousands of engineers and scientists spanning many different companies and institutions globally are now looking at that challenge.

Three factors make this an especially difficult task: first, it is hard to replace the energy capacity of fossil fuels. Second, the world population is projected to substantially grow for approximately four more decades, adding another two to three billion people to the current seven billion, about a thirty-five percent increase. Third, energy demands per person are increasing, particularly in the developing world, where lifestyles can go from primitive, off-the-grid, to modern consumers in as little as a decade.

Combined, these three factors make the challenge of reducing atmospheric carbon formidable. Many have prescribed that we simply reduce our energy demand, or that we might implement more energy efficiency in our buildings. Laudable as those efforts are, they will not solve the greenhouse gas problem, the rising temperature, and the melting ice. The important and vigorous debate about energy solutions is too complicated to thoroughly address here.

What we can briefly cover are a few technological and legislative approaches that have potential to reduce the gradually warming climate that is causing the ice to melt and the sea to rise, and open the curtain on a few that do not. Again, it is beyond the scope of this book to examine each of these in detail, but we can present an overview for those curious about possibilities. At the end of this book, I include resources for further reading.

NEW TECHNOLOGIES

Solar/Wind: Solar and wind are making incredible strides, but presently only account for a small percentage of overall energy supply, though that is changing. There are challenges with the inconsistent supply and storage requirements associated with wind and solar that will necessitate additional progress with storage and distribution. Neither are out of reach, but nor can they be implemented and scaled immediately.

If we look at the broader category of "Renewable Energy", according to the 2012 *Global Energy Assessment* as of 2009 it accounted for 17% of the global primary energy supply. Their detailed analysis indicates that renewables could reach between 20-75% of needs by 2050, and 30-95% by the year 2100. For more, see, www.globalenergyassessment.org.

Further below I describe some truly innovative concepts such as fuel from algae that hold significant potential and would obviously add to the renewable energy potential.

Nuclear: Nuclear energy has essentially zero carbon emissions, but has its own challenges. Japan's Fukushima nuclear disaster in early 2011 made these painfully clear, and ultimately put nuclear on hold for many countries, including Japan.

Understandably the unique risk and history of nuclear makes it the "third rail" for a lot of people, conjuring up associations with past accidents, nuclear weapons, and potential catastrophe.

Nonetheless it is one of the few existing technologies that can produce significant fuel without any annual additions to the level of greenhouse gases and the associated warming and melting of the ice sheets. I am also persuaded that the US nuclear Navy has had a great track record. There are new and even better potential designs for nuclear power that I believe should be 'on the table' given what is at risk from extreme climate change.

Furthermore, the Japanese nuclear disaster may prove to have a silver lining. That country is one of the most highly industrialized societies in the world, and has huge power needs. At the same time, it has minimal domestic energy sources, so is very vulnerable. With nuclear energy off the table, they are now laser-focused on sustainable alternative energy. Given their outstanding track record for technical innovation over the last four decades, their necessity very well might produce some amazing technology that benefits us all.

Some exciting nuclear technologies are also being explored, including the elusive nuclear fusion and some innovative fission reactors. While these could eliminate the big problems with the nuclear fission now used in power plants, huge technology hurdles remain. They should be researched and explored but cannot be counted on as clean energy sources at this time.

Advanced and "Clean" Coal Technologies: Coal is the most abundant fosil fuel in the United States. According to the International Energy Agency (IEA) in March 2012, coal provides approximately 40 percent of the world's electricity needs and global demand for coal is projected to grow rapidly through 2035, particularly in China and India.

"Clean Coal" is a term that has been used in recent years to describe a variety of ways to use coal as a major energy source, without producing the damaging greenhouse gas emissions. To date, hundreds of millions of dollars have been spent in pursuit of clean coal. In the last 20 years, scientists have improved coal-combustion, with technologies such as fluidized bed combustion, supercritical and ultra-supercritical boilers, and integrated gasification combined cycle systems, which more efficiently convert coal to energy.

Yet cost-effective technologies do not yet exist to produce electricity from coal without putting vast amounts of carbon into the atmosphere. Anything that suggests otherwise is grossly misleading. The budget for public relations about clean coal in the U.S. alone is $35 million a year leading many to believe the technology was within grasp. Nonetheless, given the huge current use of coal and the reserves, cleaner coal is worth further study.

Carbon Capture, Utilization, and Storage: A new technology, referred to as carbon capture, utilization and storage (CCUS), is also being explored by industry and governments. CCUS involves capturing CO_2 (carbon dioxide) at major sources (such as electric generating facilities and industrial plants), utilizing the CO_2 for a beneficial use, and/or storing the CO_2 in geologic formations, in terrestrial plants and soil, or beneath the ocean floor.

Again according to economists, CO_2 capture technologies are not currently cost effective for commercial applications due to the energy and resources required to operate the facilities, the increased cost of construction materials, and the real estate required. This could change substantially with some of the carbon pricing concepts now being considered and described later in this chapter.

Natural Carbon Sequestration: Another approach is to use the natural ability of plants to sequester carbon. All plants take up carbon dioxide as they grow. In addition to planting more trees, which has a moderate potential, researchers are looking at fertilizing ocean plants, particularly phytoplankton. This approach has real potential because it uses natural biologic processes that do not require us to generate more energy for the process. Not only do the oceans cover almost 72 percent of the planet, but they are deep, an average of two miles. Compared to the thin layers of life on land, the oceans comprise 99 percent of the total living space on earth. The tiny omnipresent algae, phytoplankton, are the biggest biomass on the planet. They have decreased by some 40 percent in the last five decades. Although there has been some controversy about this approach, a major article in July, 2012, in *Nature*, caused the idea to be revisited by the scientific community.[126] Some experiments are currently being fast-tracked to evaluate large-scale feasibility, including ocean iron fertilization. Once derided as risky, it is now warranting a fresh evaluation in the light of recent experiments, as well as the growing risk of ever-increasing levels of carbon dioxide in the atmosphere.

Ethanol: In the US a large effort has been put into corn ethanol in the last decade. Corn ethanol still creates greenhouse gases, by some estimates even more than traditional gasoline, when all factors are considered. The main argument for ethanol is around energy security, because using ethanol shifts the payment from imported sources of crude oil to an American farming-based industry. The result has been the development of a whole new industry based on billions of dollars in new subsidies, which dramatically drive up the cost of this basic food item, both in the U.S. and in other countries, while also displacing other agriculture. I am not an advocate of corn ethanol.

Natural Gas: It would be hard to have missed the energy revolution of the last few years, the use of natural gas, produced largely by the technique called "fracking" due to the way underground rock is fractured. It is both a very controversial process due to alleged damage to the aquifers, and a huge economic boon since it is inexpensive and replaces some of the US oil imports. Fracking is also expanding internationally now in many locations.

Again this book is not the place to properly cover such technology. I will confine my remarks here to its impact on warming the planet, melting the ice, and raising sea level.

Natural gas is essentially methane, one of the potent greenhosue gases covered earlier. When it is burned it produces much less carbon dioxide than coal, making it better in terms of reducing greenhouse gases, a very positive effect.

The downside is that the unburned methane that gets into the atmosphere is hundreds of times more powerful than CO_2 on a per molecule basis. Poorly done fracking leaks lots of gas. Typical transfer operations as it is moved and put into vehicles also release the gas. Recent monitoring studies confirm increasing rates of methane leakage following the dramatic use of fracking. In an OpEd in the *New York Times* on July 28, 2013, Anthony R Ingraffea, one of the engineers who pioneered the technology, describes its current use as a "Gangplank to a Warm Future."

Fuel from Algae: There is another exciting process that shows potential to be a renewable fuel source and sequester carbon, thereby improving the greenhouse gas problem from both ends.

I have heard about experiments to grow algae for conversion to fuel for several years, some undertaken by big oil companies. They never persuaded me they were economical.

Shortly before the second edition goes to press, I had the opportunity to visit the facilities of Algenol on the west coast of Florida and came away impressed and excited. Their proprietary technology appears to produce ethanol and a diesel like product from algae, cost effectively. They have an amazing lab and experimental production facility. This project is state of the art. If they are successful at scaling up to full production in 2014 it holds great promise. The process uses carbon dioxide to accelerate the algae growth, producing fuel, and sequestering carbon. For more information, look at www.algenol.com This is worth watching.

Scrubbing: Removing some of the massive carbon dioxide that we have already put into the atmosphere, or "scrubbing," has interesting potential. Scrubbing has been done for a century on a small scale and is essential for such closed systems as submarines. The challenge is to do it on a large scale in an energy efficient way. Most methods developed to date require lots of energy, which is how we got into this situation in the first place. There are some interesting possibilities using catalytic substances, perhaps on giant grids. Other researchers have created what are sometimes referred to as "artificial trees" to absorb carbon dioxide. (Some artist renderings are at: johnenglander.net/scrubbers.) The other major challenge with this concept is what to do with the carbon once it's collected. The volume of the material we need to store is immense. There are some experiments in the U.S., Norway, and in Germany to store the carbon dioxide deep underground. The early results show potential, but have limitations.

Reducing Heat Received From the Sun: This approach doesn't reduce the amount of greenhouse gases that trap heat in the atmosphere, but instead deflects the sun's rays to reduce the incoming heat. Concepts range from using a rocket that would spread tiny reflecting pieces into space to spraying clouds with seawater to make them more reflective. Both are under study.

It would be premature to count on any of the actions described above, but they are a source of some hope that we may be able to slow the pace of atmospheric warming. These technologies are changing on a monthly basis and receiving considerable support from industry, governments, and foundations in various countries. You will undoubtedly see more media coverage and other sources of information on each of them.

It is likely that sooner or later we will have to try a technological approach to reverse the effects of a warming atmosphere. At the same time, we must keep in mind that intentionally changing earth systems can bring unintended consequences.

Geo-Engineering - A Faustian Bargain?

The techniques just above that actively remove carbon dioxide fro the atmosphere, or reduce the amount of sunlight received are generally described as "geo-engineering" meaning that humans are altering Earth's climate. A large portion of the scientific community thinks geo-engineering is just too risky ever to be justified.

It seems to me that with our population at seven billion headed to ten, and by altering the mix of the atmospheric gases so that we are warming rather than cooling, that we crossed the line over to geo-engineering some time ago, even if it was not intentional. As climate gets really inhospitable, I see little chance that we will not resort to some form of geo-engineering. It should be done with the greatest possible caution.

Some of these concepts that have been floated around are simply not currently realistic. For example, some have suggested we disperse sulfur dioxide into the stratosphere, which theoretically would deflect some heat and cool things off. In their best selling book, *Super Freakonomics: Global Cooling, Patriotic Prostitutes, and Why Suicide Bombers Should Buy Life Insurance,* Steven D. Levitt and Stephen J. Dubner tout the economics of cooling the planet with such a technology, without understanding the side effects. This concept has been dismissed by anyone taking the issue seriously. Adding sulphur dioxide would bring back high levels of "acid rain," along with the many negative health impacts that led to the creation of the Clean Air Act in the 1970s. It would also greatly worsen ocean acidification, a serious problem described in Chapter 11 that could accelerate changing the basic chemistory of our ecosystem. In effect, the sulfur dioxide cure could be worse than the illness.

Political Pandering

Today, environmental legislation has a bad image to some who believe there has been too much of it. While this may be true in certain circumstances, few would dispute the need for basic protection of the clean air and water that we all require. These both are sometimes referred to as "the commons" and may include public lands, rivers, and oceans.

The Clean Air Act and Clean Water Act have proven themselves essential to the quality of life we enjoy in the United States. When enacted, these pieces of legislation were not considered partisan concerns, and in fact were passed during the Republican Nixon administration, in recognition of the need to protect human health and welfare. Given the vast impact of climate change affecting all of us, we face a situation that requires the same non-partisan understanding today.

To date, the U.S. Congress has been disturbingly close-minded to energy initiatives that address climate change. In 2011, Congress thwarted efforts to implement carbon trading programs or any other mechanism to put a cost on emissions. Nonsensical arguments included carbon dioxide being a natural substance and the fact that cows fart methane.

Indeed, the politicization of climate concerns over the last decade is one of the most unfortunate and bizarre twists of this difficult issue. This political reality was best demonstrated when all the Republican candidates for U.S. president in 2012 disputed the scientific finding about greenhouse gases and climate change (with the exception of Jon Huntsman, one of the early contenders). *The New York Times* estimated that 153 million dollars worth of ads had been placed just during the 2012 U.S. presidential contest in support of fossil fuels and criticizing clean energy. The Republican leadership has made science in general, and climate change in particular, a target for suspicion and derision. That position about science is bizarre and embarassing. Some suggest that this unrealistic position contributed to the Republicans loss in the 2012 election.

But there is enough blame to share on both sides of the political aisle. As soon as the Democrats got control of the House of Representatives in 2007, they enacted The Energy Independence and Security Act of 2007 (originally named The Clean Energy Act of 2007). One of its highly touted goals was to get corn ethanol production scaled up in this country. As described, ethanol may produce even more greenhouse gases than traditional gasoline and presents a host of other problems. If that was the best shot the Democrats had to demonstrate their climate change priorities, they failed miserably.

Setting aside the entire policy of subsidies and the huge lobbying dollars that influenced this legislation, it is disturbing that the whole effort was cloaked as "clean energy," implying that it would begin to address the challenge of climate change. Many Washington insiders recognized this was a contest between the agricultural sector and the petroleum sector. We need real solutions, driven by the public interest, not by the group that spends the most to influence legislation. It is no wonder the public holds Congress in such low esteem.

Those who put their political party above reality, who "spin" this issue, or continue with blind loyalty, should look at their children or grandchildren and have a moment of reflection. They can also think about their personal legacy. Recall that it took three or four decades for some of the staunchest defenders of racial segregation to see the light and to recant their obstinacy. No doubt this issue will see its version of the late Senator Strom Thurmond and Governor George Wallace, just to name two of the most ardent segregationists who late in life realized, deeply regretted, and admitted their mistake, which was born from bias and maintained by stubborness.

Fortunately, there is some good bi-partisan leadership emerging on the issue. For example, former South Carolina Congressman Bob Inglis started a new initiative, the Energy & Enterprise Institute, which exemplifies a fresh conservative concern about the core issue of energy technology. A surprising number of conservative, and even religious groups, now recognize the long-term threat posed by climate change to future generations, to the less fortunate, and to the majesty of the natural world. Some of these include the Young Conservatives for Energy Reform, Evangelical Environment Network, Resources for the Future, Catholic Climate Covenant, and Christians and Climate.

The environmental field is often thought of as liberal. But there are many groups, including hunters, that are very concerned about conservation of the natural world. None has been more articulate about this threat than Larry Schweiger, president of the huge conservation organization, National Wildlife Federation. (See his small but excellent book, *Last Chance: Preserving Life on Earth.*) Many others could be cited.

As I have noted, the military will likely play a leadership role on this issue. As described in Chapter 11, methodical analysis and planning is one of its prime responsibilities. And when senior military officials talk about a clear and increasing danger and the need to prepare and adapt to sea level rise, people are sobered.

We have quite a track record of preferring to avoid the painful truth that stares us in the face. Leon Panetta, now Secretary of Defense, and previously Director of the U.S. Central Intelligence Agency (C.I.A.) often remarks, "We can govern by leadership or we can govern by crisis. Without leadership, we will continue to find ourselves governing by crisis." These are apt words for the choices we face in terms of dealing with the profound change caused by ever-rising sea level.

Legislation with "teeth" will be required to address climate change and sea level rise. Somewhat like the levels of national debt, delay greatly compounds the problem. One of the easiest ways to change societal behavior is through financial incentives.

"Cap and Trade" proposals have been tossed around for years. This approach would allocate greenhouse gas emissions among various industries and businesses. If they exceed their quota, they would have to buy (trade) an offsetting amount of carbon. They could buy from a company that has decreased its emissions, or a firm that reduces carbon, perhaps via tree planting. While Cap and Trade has gained favor in the last decade as a market-based mechanism, it has its critics, vulnerabilities, and has yet to prove that it can achieve its goals.

A **carbon fee or tax** is just as unpopular as any tax. Yet many industrial leaders, including some in the oil industry, have come to see that it may be an effective pricing mechanism that can work on a long-term, escalating cost basis so that industry can plan accordingly. The fee structure does not need to be an additional tax. A carbon tax could even replace existing ones such as income, sales, or excise taxes.

In many places the cost of beverage cans, plastic bottles, batteries, or automobile tires includes a fee to cover the cost of disposal and recovery. While it is an intrusion on our lives and laissez-faire, many of us can accept this market-based way to reduce litter and recycle by means of a financial incentive, rather than the brute force of a law with punitive fines.

Fee and Green Check: Recently, Dr. James Hansen stepped beyond his role as leading climate scientist into the policy arena to advocate a pricing mechanism known as "Fee and Dividend" or "Fee and Green Check."

The idea is to add an escalating fee to energy sources based upon the amount of carbon they put into the atmosphere. The fees would not be kept by government, but would be completely distributed back to taxpayers as a dividend, or "green check." Whether it's politically possible is a fair question. Regardless, it's a fascinating market-based concept.

The point is that there are some mechanisms that could motivate us all to drastically reduce greenhouse gases over the course of this century. Legislative support for an energy policy that puts a price on carbon increases would translate a known danger into a monetary formula so that the private sector can adjust creatively. But first, Congress and other legislative bodies need to embrace the concept, setting aside the short-term interests of some industries. In the Resources section of this book and on the website (www.johnenglander.net) I list some organizations that are pursuing such community and society-level efforts.

The Business Community Is Not The Enemy

While a few companies spin nonsense, corporations can be important vehicles for change, mobilizing vast resources, and responding quickly and with creativity. It is up to the press, government, and private citizens in diverse communities to act as watchdogs, identifying cases where particular companies damage our common entitlement to sustainable air, water, and oceans.

Healthy partnerships between government and businesses will be essential to our ability to adapt. This can happen when government sets and implements clear energy policies and invests in basic technology, and then allows businesses to develop products and services in a free and competitive market. I simply do not see any other way that we will make the massive changes that we need in time.

People are getting tired of polarization. We need to find a better way. The challenge of rising sea level will be as good a test as any to see if we can change the way we govern ourselves. Government can bring brute force and create the framework, but generally speaking, companies are much more nimble.

I believe that free markets are the most flexible and effective vehicles for change. While some are bound to the past, there are dozens of major corporations including many in the energy field that simply want to know what the pricing mechanisms will be for carbon in the decades ahead so that they can plan and move forward. They are still waiting.

Chapter Seventeen
Moving to Higher Ground

Four hundred years ago Galileo ran into the dogma of the Catholic Church because he dared voice the radical concept that the sun is the center of our planetary system (heliocentrism). His trial by the Inquisition and the ultimate guilty verdict for heresy left him under house arrest for the rest of his life. That story still stands as a powerful example of the clash between science and cultural beliefs.

In our time, science has again come under attack, but this time not by the Church. In fact, though largely overlooked, the Vatican has taken an exemplary position regarding the catastrophic challenge posed by climate change, and specifically by sea level rise. In May 2011, the Pontifical Academy of Science produced an easily read, 17-page, outstanding report[127] (download at johnenglander.net/vatican). Its summary, three major findings, and powerful one-sentence conclusion are:

> *Since a sustainable future based on the continued extraction of coal, oil and gas in the "business-as-usual mode" will not be possible because of both resource depletion and environmental damages (as caused, e.g. by dangerous sea level rise) we urge our societies to:*
>
> *I. Reduce worldwide carbon dioxide emissions without delay, using all means possible to meet ambitious international global*

warming targets and ensure the long-term stability of our climate system.

II. *Reduce the concentrations of warming air pollutants (dark soot, methane, lower atmosphere ozone, and hydrofluorocarbons) to slow down climate change this century.*

III. *Prepare to adapt to the climatic changes, both chronic and abrupt, that society will be unable to mitigate.*

The cost of the three recommended measures pales in comparison to the price the world will pay if we fail to act now.

~From *Fate of Mountain Glaciers in the Anthropocene*

The case could not be made more succinctly or firmly, and stands in sharp contrast to the explicit efforts of the U.S. Congress to squash measures to address the problem and deny the realities we face.

OUR BEST SHOT

For almost 12,000 years humans have lived in an epoch known as the Holocene. That designation will likely soon change. In 2011 scientists began a process to officially declare this new time period the "Anthropocene," or human-forced era. New epochs are used to define dramatic changes in climate typically evidenced by major alterations to plant and animal life.

As cited, the last time earth had similar temperatures to the present, 120,000 years ago, sea level was 26 feet (six meters) higher than today, possibly more.[18] Even then, carbon dioxide levels never exceeded 280 ppm. Now they are nearing 400 ppm and headed higher. "Business as usual," burning coal and extra "dirty" versions of petroleum, like the tar sands, is projected to produce carbon dioxide levels of 700-900 ppm by the end of the century, which would heat things up significantly and quickly.[67]

On the current trajectory, over the next century or two, the ocean could well rise as much as 20 feet and higher. It is doubtful whether it is even possible over the next century to prevent ice sheets from reaching catastrophic "tipping points."

In addition to the urgency of intelligent adaptation, I believe our primary goal must be to keep the level of carbon dioxide from going above 450 ppm for any extended period of time, and to return to 350 ppm as soon as possible.

Our best chance as individuals and as communities to reach this goal is to demand that our political leaders support broad systems initiatives with real potential to reduce the level of atmospheric carbon dioxide.

Of course, this is an international issue, and solutions must be agreed upon by the U.S., China, India, and Europe as the four prominent contributors. Behind the political posturing and accusations, those with knowledge of the international climate talks say that China will take a strong position to restrict carbon dioxide emissions if the U.S. is willing to show leadership.

We must push our leaders to focus on overhauling our energy systems, incorporating energy efficiency into buildings, and putting the maximum effort into the research and development of low-carbon energy distribution and storage systems. Some work is proceeding on all of these initiatives, but we must take on a much greater, more concerted effort.

"The next generation will have to figure this out" is something I hear often. This is simply an escape, or cop-out. Waiting 20 or 30 years for a new solution that will then take the usual couple of decades to implement is either uninformed or insincere, and those are the polite terms. By then it will be too late to prevent catastrophe, due to the various lag times and the amount of time it takes to fully implement a new technology, particularly one that could generate a substantial portion of our power supply. There is also the lag time for a cooler atmosphere to cool the ocean and stop thermal expansion and melting of the ice sheets.

If you say "politics is not for me," you are missing your best opportunity. Whether we like politics or not, we need to work with it to achieve this important goal. This issue, like clean air and clean water, requires government regulation, or at least a long-term operating framework in which the private sector can work competitively.

Questions about policy and planning for sea level rise have recently been raised all along the coast from Boston to New York City, New Jersey, Delaware, Virginia, the Carolinas, Florida, around the Gulf of Mexico, up the Pacific Coast and in communities as far flung as Australia and Bangladesh. In diverse communities across America and around the world, informed people are connecting the dots and using some common sense. Even as this goes to press, a stunning 72 percent of the American public recognizes the looming climate crisis.[128] This is in sharp contrast to the political impasse at the national level.

Organizations have started to form. The list of such groups is changing and expanding all the time. On my website I will try to keep a resource page called "Community" where I will list them. Please let me know of groups that should be added. Also check for your local chapter of 350.org, which was started by Dr. Bill McKibben, at that organization's website: www.350.org. This organization is leading the international grassroots effort to reverse the growth in greenhouse gases.

Another organization that I can recommend is Citizens Climate Lobby, working to encourage political support on the issue. Their website is: www.citizensclimatelobby.org.

(There are some other interesting efforts that are not quite ready to announce. If you would like to be kept informed please look at my website, www.johnenglander.net)

Recycling And Other "Green" Actions?

At the risk of taking on a sensitive topic, many people feel that personal accountability in addressing this issue starts and ends with "green" actions such as recycling.

For the record, my family owns a hybrid car, has swapped out our light bulbs, installed a solar water heater, and sorts our trash for recycling. All of these things make a tiny, perhaps microscopic, difference, but more importantly demonstrate a level of commitment and personal responsibility. Like many of you, we feel good by doing things that are "green" and sustainable.

While it upsets many to hear this, the unspoken truth is that while recycling is a good concept and can conserve resources, it does little if anything to address the problem of climate change or sea level rise. The key is to distinguish the conservation of *things* from the conservation of *energy*. A lot of recycling does not conserve energy, which we must do to reduce greenhouse gases. Certain kinds of recycling are very energy inefficient; that is, they use more energy than would be used creating a new product.

Even if everyone recycles everything possible, the planet will continue to heat up and the ice sheets will continue to melt. In fact, changing home thermostats by a degree or two saves more energy than recycling. Modest energy reductions from recycling are outweighed by continued huge population growth and increased energy demand around the world.

The point is that it's imperative not to let relatively easy actions become a feel-good exercise, possibly distracting us from things that could make a real difference in slowing climate change and rising sea level. We need a concentrated, long-term effort to produce energy without producing carbon. This can be a job-creating effort with long-term economic benefits. The effort required may be even larger than the scale of the space program, the fight against cancer, or the resources put into the war against terror. Not that any of those things are not important, but the long-term change to our climate, the melting of the ice sheets, and the steady rise of sea level surely warrants our maximum effort and focus.

Adaptation Is Inevitable

While personal and political actions are key, we must recognize that we'll face enormous psychological resistance to accepting this future scenario. The human need for connection to place, a sense of security and a sense of predictability are all at stake in this discussion, and any idea that threatens those needs won't be readily accepted.

Other global upheavals, such as world financial crises, terrorism, disease, floods, fires, droughts, famine, water supply, or national security may also continue to delay attention to the long-term threat of sea level rise.

Most national, state and local governments already have badly strained budgets and high debt loads, making them all the more likely to deal with near-term crises and necessities rather than long-term planning. And while some locations are leading the way in starting to plan for significant rising sea level, funding the implementation of ambitious long-term visionary projects is daunting at best in the face of budget demands for police, social programs, health care, schools, roads, bridges, and adequate supplies of energy, water, and food.

Despite all this, at some point the increasingly immediate consequences relating to climate change and sea level rise will make it a priority. A series of storm surge events might bring attention to vulnerabilities of coastal areas and jolt us into confronting the new coastal reality. We might find a political leader, possibly a future U.S. president, who will use that position to make a compelling case for preparation, and maybe even set visionary policy. Likely, that will be several years in the future, perhaps in the 2020's, when rising sea level will be impossible to ignore.

Over the next few decades, many vulnerable coastal cities will somehow find the hundreds of millions of dollars needed for civil engineering that can buy a few decades of survival. If their geology supports a fix for, say another 50 years, the investment probably makes sense and will happen. True survival has an amazing power to help people set priorities.

At some point—later this century for many—most coastal communities will get to a point where their outlook for rising waters, increased storms, and underlying geologic structure will preclude further massive investment in community infrastructure. Each community's options will vary based on elevation, geologic structure, population density, and wealth. Manhattan, Miami, the Maldives, Manila, and Monaco—just for alliteration that spans the globe—all face different challenges and will find different solutions.

There will be those who see it coming and take steps to leave before things deteriorate. Before anyone notices or admits it, the population and real estate prices will decline just a bit. Once noticed, the trend will accelerate. Others will hang on until the last possible day, as happened at Sharps and Holland Islands in Chesapeake Bay.

Governments will see things degrade over decades and, eventually, year-to-year. Disputes will probably range from how bold and farsighted the adaptation should be to a focus on who should bear the burden of paying for the damages, the adaptation, or relocation.

Our coastal cities will not slip beneath the waves as innocently as those islands in the Chesapeake Bay. It will be a treacherous shoreline as our huge buildings and infrastructure rust and collapse. Sooner or later those cities will become shantytowns and outposts, reminiscent of old, ghost towns in the American West, to be explored in the next century by scuba divers.

Cancer: A Good Metaphor

It usually catches people off guard when I say that cancer is a good metaphor for sea level rise and climate change. Most of us have been touched by that deadly disease, whether it has hit us personally, or has struck a family member or a friend.

As recently as the 1960's a diagnosis of cancer was little different than being told you were going to die rather soon. Though still depressing and scary, that diagnosis today is a different story. For a couple of days, we might freeze like a deer in the headlights. We might cry or just go home and not be able to get out of bed.

After processing the news, inevitably we research our options and begin to deal with it. Fortunately, the outcomes are vastly improved. For most, cancer is no longer a death sentence. Indeed, today we are far more likely to know of cancer survivors than those who lost the battle.

Mankind put significant resources into combating the cancer problem and has had very good results so far, with more in the pipeline. I have to believe that if we put the same kind of effort into finding and implementing low carbon energy sources we can slow the threat of sea level rise too and begin intelligent adaptation.

Closing Thoughts

The more we know about the scale of sea level rise in the decades ahead, the better we can plan for it. The new era of shifting shorelines has begun and will be much more obvious by mid-century. It demands honest discussion and practical long-term community planning.

Rising seas and erosion will challenge us to do some extraordinary civil engineering, far beyond simple seawalls and levees. With the right attitude and resources, we will probably come up with some very clever ways to reduce some of the damage through worthwhile investments.

We have the opportunity to make plans and lay foundations for structures that will have value for centuries. It's time to create a vision. This will require a profound change in attitude about where the coastline will be in the coming decades and centuries. It will be a slowly moving and meandering target, rather than a fixed line. It challenges us to change our concepts about real estate and permanence.

We might take inspiration from those who, many centuries ago, built and invested in the great cathedrals that are hallmarks of architecture, construction, and generational legacy. These took generations, even centuries, to build. People devoted their lives and wealth to the endeavor, knowing that they would not personally enjoy the fruits of their efforts. The beneficiaries would be their children, grandchildren, and beyond. Dealing with sea level rise is going to require some of the same multi-generational perspective.

Now that you understand the phenomenon of rising sea level, you have the opportunity to spread awareness, advance efforts towards intelligent adaptation, and begin to protect your financial exposure. There may not be a more important gift to future generations. Consider what will happen on our current path of business-as-usual if we deny, delay, or are incapable of action on this pivotal issue for our civilization.

Alternatively, one can at least imagine how this issue could challenge, galvanize, and inspire us. Possibly, it could give us an opportunity to grasp what is truly important at home, in our communities, our country, and on our planet.

One of the most inspiring quotes of history was Sir Winston Churchill exhorting England to do what was necessary to prevail against the

Third Reich. It is surprisingly fitting to this issue and context:

> *If we fail, then the whole world… including all that we have known and cared for, will sink into the abyss of a new Dark Age.... Let us therefore brace ourselves to our duties, and so bear ourselves that, if [we*] last for a thousand years, men will still say, 'This was their finest hour.'*[129]
>
> * "we" replaces the British Commonwealth and Empire

Could this battle be our finest hour? Do enough people see where we are headed into the next century? Or are we like a ship meandering along a river, unknowingly headed for a great waterfall? Add to that image the different groups of passengers on board that ship, some partying, others arguing and fighting, and some doing business, but few looking ahead.

Could the reality of rising sea level, which will surely intersect with our valuable coastal real estate, be a vehicle to change course? I do not know the answer to that any better than you. What I do hope is that the facts I have assembled and shared allow you to be one more person who can consider that possibility in its many variations.

To close, I will share a vignette that comes back to me from childhood. It merely serves as a metaphor to put in perspective the world we knew decades ago and the entirely different one that we now see on the horizon.

My childhood summers were spent on Martha's Vineyard. My hangout was the fishing dock in picturesque Menemsha, dominated by commercial vessels bringing in swordfish, lobster, and shellfish. The boats at the wharf moved up and down with the changing tide. At full moon or during storms the water might come over the massive docks, but just barely.

As I think about the changing world, a particular afternoon there in the 1960's comes to mind. It was a typical summer day. An old fisherman, smoking a pipe, worked to repair his nets, getting his gear in shape for the next voyage. A man and woman got off a tour bus and walked over to the seaman, who would not be distracted from his work. To engage the local, and I suspect to impress the lady, the man asked, "Is the tide coming in, captain?" Without stopping or even lifting his head to make eye contact, the gravelly New England voice responded, "Likely. Always has." With just a hint of sarcasm, his terse reply described the predictable rhythms and pattern of the sea, the certainty of his world and the generations before his.

Even though the effects may not be seen until that fisherman's great grandchildren's generation, we all must recognize a new certainty: our coastal civilization will be totally transformed. That safe harbor will disappear or be rebuilt at a higher elevation as the tide level continues to advance. In terms of practical impact, this new perspective of rising sea levels may have even more impact than Galileo's.

Like the responsible sea captain, we should begin to prepare our ship for the voyage ahead.

Afterword

What can we each do? That is the inevitable question. The subject of rising sea level is sobering and scary. Hopefully, you are compelled towards positive action. In the event that my personal direction provides any example for you, I will share three different avenues.

First, make a plan for your personal investments, particularly coastal real estate and other investments that might be affected negatively or positively by long-term sea level rise.

Second, share what you have learned with your local community and your broader network. Obviously, this book is part of that process for me. It is one component of my work to help spread the information and how it can benefit individuals, businesses, and communities.

Third, turn your knowledge, perspective and concern into support for political leaders who "get it." Several organizations are focusing on this, but it is an early effort, needing much more support.

If you have other ideas, please share them with me. Perhaps I can help get them to a wider audience. This is such an enormous challenge that lots of efforts are needed. If you would like to join my network, this book has a Facebook page "High Tide On Main Street." You can also sign up on my website for the latest information: www.johnenglander.net.

Thank you.

John Englander

john@johnenglander.net

Resources and Links

Websites:

- climate.gov
- skepticalscience.com
- climateprogress.com
- epa.gov/climatechange/
- 350.org
- climatecentral.org
- nsidc.gov
- climatescoreboard.org
- citizensclimatelobby.org

Further Reading:

Plan B 4.0: Mobilizing to Save Civilization, by Lester Brown. New York: W.W. Norton, 2009.

Earth, the Sequel: The Race to Reinvent Energy and Stop Global Warming, by Fred Krupp and Miriam Horn. New York: W.W. Norton, 2008.

Heat: How to Stop the Planet from Burning, George Monbiot. Cambridge, MA: South End, 2007.

Last Chance: Preserving Life on Earth, by Larry J. Schweiger. Golden, CO: Fulcrum Pub., 2009.

The Global Deal: Climate Change and the Creation of a New Era of

Progress and Prosperity, by N.H. Stern. New York, NY: PublicAffairs, 2009.

World on the Edge: How to Prevent Environmental and Economic Collapse, by Lester Brown. New York: W.W. Norton, 2011.

The Weather of the Future: Heat Waves, Extreme Storms, and Other Scenes from a Climate-changed Planet, by Heidi Cullen. New York: HarperCollins, 2010.

The Post Carbon Reader: Managing the 21st Century's Sustainability Crises, by Richard Heinberg and Daniel Lerch. Healdsburg, CA: Watershed Media, 2010.

The Quest: Energy, Security and the Remaking of the Modern World, by Daniel Yergin. New York: Penguin, 2011.

For Further Information:

See the author's website, for his blog, to sign up for emails, and to contact him: www.johnenglander.net

References

1. **Allison, I. et al. (2009).** Ice sheet mass balance and sea level. *Antarctic Science*, 21(05), 413-426. Retrieved from http://journals.cambridge.org/abstract_S0954102009990137

2. **Kumar, P. et al. (2007).** The rapid drift of the Indian tectonic plate. *Nature-London*, 449(7164), 894. Retrieved from http://www.eas.slu.edu/People/DJCrossley/gjc/talks09/john_nature06214.pdf

3. **Kent, D. V., & Muttoni, G. (2008).** Equatorial convergence of India and early Cenozoic climate trends. *Proceedings of the National Academy of Sciences*, 105(42), 16065. Retrieved from http://www.pnas.org/content/105/42/16065.short

4. **Macdonald, F. A. et al. (2010).** Calibrating the Cryogenian. *Science*, 327(5970), 1241. Retrieved from http://www.sciencemag.org/cgi/content/abstract/sci;327/5970/1241

5. **Bice, K. L. et al. (2006).** A multiple proxy and model study of Cretaceous upper ocean temperatures and atmospheric CO2 concentrations. *American Geophysical Union*. Retrieved from http://darchive.mblwhoilibrary.org:8080/handle/1912/846

6. **Urrutia-Fucugauchi, J., & Camargo-Zanoguera..., A. (2011).** Discovery and focused study of the Chicxulub impact crater. *Eos Trans. AGU*,. Retrieved from http://www.agu.org/pubs/crossref/2011/2011EO250001.shtml

7. **Alvarez, L. W. et al. (1980).** Extraterrestrial cause for the Cretaceous-Tertiary extinction. *Science*, 208(4448), 1095.

8. **Ruhl, M. et al. (2011).** Atmospheric Carbon Injection Linked to End-Triassic Mass Extinction. *Science*, 333(6041), 430-434.

9. **Shellito, C. J. et al. (2003).** Climate model sensitivity to atmospheric CO2 levels in the Early-Middle Paleogene. *Palaeogeography, Palaeoclimatology, Palaeoecology*, 193(1), 113-123.

10. **McPhee, J. (2000).** *Annals of the Former World*. Macmillan.

11. **Grifantini, K. (2011).** Where Did Earth's Water Come From? *Sky and Telescope*, p. 22.

12. **Schmidt, G. (2010).** [Dr. Schmidt is a highly regarded NASA climate modeler] Estimate of Milankovitch forcing variation. Personal communication.

13. **Alley, R. B. et al. (2003).** Abrupt climate change. *Science*, 299(5615), 2005.

14. **Team, NASA-GI.S.** Panama: Isthmus that Changed the World. Retrieved from http://earthobservatory.nasa.gov/IOTD/view.php?id=4073

15. **Cronin, T. M., & Dowsett, H. J. (1996).** Biotic and oceanographic response to the Pliocene closing of the Central American Isthmus. *Evolution and Environment of Tropical America,* 76-104. Retrieved from http://books.google.com

16. **Peltier, W. R. (1994).** Ice age paleotopography. *Science*, 265(5169), 195–201.

17. **Hancock, G. (2002).** *Underworld: The Mysterious Origins of Civilization.* Crown. Retrieved from http://www.amazon.com/Underworld-Mysterious-Origins-Civilization-

18. **Kopp, R. E. et al. (2009/12/17/print).** Probabilistic assessment of sea level during the last interglacial stage. *Nature*, 462(7275), 863-867.

19. **Fedje, D. W., & Mathewes, R. W. (2006-07-30).** *Haida Gwaii: Human History And Environment from the Time of Loon to the Time of the Iron People* (Pacific Rim Archeaology). Univ of British Columbia Press.

20. **Reid, M. (2008).** *Myths & Legends of the Haida Indians of the Northwest: The Children of the Raven.* Santa Barbara: Bellerophon Books.

21. **Ruddiman, W. F. (2010).** A Paleoclimatic Enigma? *Science*, 328(5980), 838.

22. **Kaufman, D. S. et al. (2009).** Recent warming reverses long-term arctic cooling. *Science*, 325(5945), 1236.

23. **Tripati, A. K. et al. (2009).** Coupling of CO2 and ice sheet stability over major climate transitions of the last 20 Million years. *Science*, 326(5958), 1394.

24. **Rohling, E. J. et al. (2007).** High rates of sea-level rise during the last interglacial period. *Nature Geoscience,* 1(1), 38-42.

25. **Hansen, J., & Sato, M. (2011)**. Paleoclimate Implications for Human-Made Climate Change. *Climate Change: Inferences from Paleoclimate and Regional Aspects,* 21-48.

26. **Blanchon, P. et al. (2009)**. Rapid sea-level rise and reef back-stepping at the close of the last interglacial highstand. *Nature,* 458(7240), 881-884. Retrieved from

27. **Tol, R. S. J. et al. (2006)**. Adaptation to five metres of sea level rise. *Journal of Risk Research,* 9(5), 467-482. Retrieved from http://fnu.zmaw.de/fileadmin/fnu-files/publication/tol/jrrslr5.pdf

28. **Will, G. (11-26-2010)**. The Earth Doesn't Care. *Newsweek,* 26. Retrieved from http://www.newsweek.com/2010/09/12/george-will-earth-doesn-t-care-what-is-done-to-it.print.html

29. **Domingues, C. M. et al. (2008)**. Improved estimates of upper-ocean warming and multi-decadal sea-level rise. *Nature,* 453(7198), 1090-1093.

30. **Church, J. A. et al. (2010)**. *Understanding Sea-Level Rise and Variability.* Oxford: Wiley-Blackwell.

31. **Meehl, G. A. et al. (2007)**. Global climate projections in Climate Change 2007: the physical science basis: contribution of Working Group I to the Fourth Assessment Report of the Intergovernmental Panel on Climate Change.

32. **Joughin, I. et al. (2004)**. Large fluctuations in speed on Greenland's Jakobshavn Isbrae glacier. *Nature,* 432(7017), 608-610. Retrieved from http://www.nasa.gov/pdf/121650main_Joughin_Nature.pdf

33. **Vaughan, D. G. et al. (2003)**. Recent rapid regional climate warming on the Antarctic Peninsula. *Climatic Change,* 60(3), 243-274. Retrieved from http://www.springerlink.com/index/U52N45201T383M4R.pdf

34. **USGS. (2-22-2010)**. Ice Shelves Disappearing on Antarctic Peninsula. Retrieved from http://www.usgs.gov/newsroom/article.asp?ID=2409

35. **Mercer, J. H., (1978)**. West Antarctic ice sheet and CO2 greenhouse effect: a threat of disaster. *Nature,* 271(5643), 321-325.

36. **Rignot, E. (2006)**. Changes in ice dynamics and mass balance of the Antarctic ice sheet. *Philosophical Transactions of the Royal Society A: Mathematical, Physical and Engineering Sciences,* 364(1844), 1637-1655.

37. **Bamber, J. L. et al. (2009)**. Reassessment of the potential sea-level rise from a collapse of the West Antarctic Ice Sheet. *Science*, 324(5929), 901.

38. **Vaughan, D. G. (2008)**. West Antarctic Ice Sheet collapse–the fall and rise of a paradigm. *Climatic Change*, 91(1), 65-79. Retrieved from http://www.springerlink.com/index/6800156543X9126J.pdf

39. **Kerr, R. A. (2009)**. Arctic summer sea ice could vanish soon but not suddenly. *Science*, 323(5922), 1655.

40. **Wang, M., & Overland, J. E. (2009)**. A sea ice free summer Arctic within 30 years? *Geophysical Research Letters*, 36(7), L07502.

41. **Vidal, J. (9-17-2012)**. Arctic expert predicts final collapse of sea ice within four years. *The Guardian*, Retrieved from http://www.guardian.co.uk/environment/2012/sep/17/arctic-collapse-sea-ice?newsfeed=true.

42. **Krupnik, I., & Jolly, D. (2002)**. *The Earth Is Faster Now: Indigenous Observations of Arctic Environmental Change. Frontiers in Polar Social Science.* Arctic Research Consortium of the United States.

43. **NSIDC.** National Snow and Ice Data Center web site. Retrieved http://nsidc.org/cryosphere/seaice/processes/albedo.html

44. **Leahy, S. (9-20-2010)**. Arctic Ice in Death Spiral. Retrieved from http://ipsnews.net/news.asp?idnews=52896

45. **Archer, D., & Brovkin, V. (2008)**. The millennial atmospheric lifetime of anthropogenic CO^2. *Climatic Change*, 90(3), 283-297. Retrieved from http://www.springerlink.com/index/T1265R6548477378.pdf

46. **(2008)**. *Global glacier changes: facts and figures.* Retrieved from http://www.grid.unep.ch/glaciers/

47. **Hall, M. H. P., & Fagre, D. B. (2003)**. Modeled climate-induced glacier change in Glacier National Park, 1850–2100. *BioScience*, 53(2), 131-140. Retrieved from http://caliber.ucpress.net/doi/full/10.1641

48. **Minard, A. (2009)**. No More Glaciers in Glacier National Park by 2020? *National Geographic News, 2.*

49. **Bradley, R. S. et al. (2006)**. Threats to Water Supplies in the Tropical Andes. *Science*, 312(5781), 1755-1756.

50. **Campbell, K. M. (2007).** The age of consequences: the foreign policy and national security implications of global climate change. Retrieved from http://www.dtic.mil/cgi-bin/GetTRDoc?Location=U2&doc=GetTRDoc.pdf&AD=ADA473826

51. **NCDC. Global Warming** - FAQ's. Retrieved from http://www.ncdc.noaa.gov/cmb-faq/globalwarming.html

52. **Bindoff, N. et al. (2007).** Observations: Oceanic climate change and sea level. Chapter Five in Climate Change 2007: The Physical Science Basis. Contribution of Working Group 1 to the Fourth Assessment Report of the Intergovernmental Panel on Climate Change, 392.

53. **Sabine, C. L. et al. (2004/07/16).** The Oceanic Sink for Anthropogenic CO2. *Science*, 305(5682), 367-371.

54. **Lean, J. L., & Rind, D. H. (2008).** How natural and anthropogenic influences alter global and regional surface temperatures: 1889 to 2006. *Geophys. Res. Lett*, 35, L18701

55. **Gerlach, T. (2011).** Volcanic versus anthropogenic carbon dioxide. *Eos Trans. AGU*. Retrieved from http://www.agu.org/pubs/eos-news/supplements/2011/gerlach_92_24.shtml

56. **McCormick, M. P. et al. (1995).** Atmospheric effects of the Mt. Pinatubo eruption. *Nature*, 373(6513), 399-404.

57. **Stommel, H., & Stommel, E. (1983).** *Volcano weather: the story of 1816, the year without a summer*. Seven Seas Press.

58. **Post, N. Y. (2-14-10).** Trump Cool To Global Warming. *New York Post, p. 6.*

59. **Maibach, E. et al. (6-30-11).** A National Survey of News Directors About Climate Change: Preliminary Findings.

60. **Homans, C. (Jan 2010).** Hot Air - Why don't TV weathermen believe in climate change? *Columbia Journalism Review*, p. Cover Story.

61. **Dahe, Q. (2013).** Climate Change 2013: the physical science basis: contribution of Working Group I to the Fifth Assessment Report of the Intergovernmental Panel on Climate Change. Cambridge Univ Pr.

62. **Dolan, K. (3-30-2009).** Physics for the Rest of Us. Forbes, http://www.forbes.com/forbes/2009/0330/030-ideas-opinions.html.

63. **WSJ (2-5-2007).** Climate of Opinion. The Wall Street Journal, Document3.

64. **Lomborg, B. (10-7-2007).** Chill Out. Stop fighting over global warming - here's the smart way to attack it. *The Washington Post.* Retrieved from http://www.washingtonpost.com/wp-dyn/content/article/2007/10/05/AR2007100501676.html

65. **Kunzig, R. & Broecker, W. (2009-01-01).** *Fixing climate: the story of climate science - and how to stop global warming.* Green Profile.

66. **Goelzer, H. et al. (2012).** Millennial total sea-level commitments projected with the earth system model of intermediate complexity LOVECLIM. *Environmental Research Letters,* 7(4), 045401. Retrieved from http://stacks.iop.org/1748-9326/7/i=4/a=045401

67. **Sokolov, A. P. et al. (2009).** Probabilistic forecast for 21st century climate based on uncertainties in emissions (without policy) and climate parameters. Retrieved from http://dspace.mit.edu/handle/1721.1/44627

68. **Rignot, E. et al. (2011/03/04).** Acceleration of the contribution of the Greenland and Antarctic ice sheets to sea level rise. *Geophys. Res. Lett.,* 38(5), L05503.

69. **Velicogna..., I. (2009).** Above-linear increase in Greenland and Antarctica ice mass loss from GRACE and other data. *AGU Fall Meeting Abstracts.* Retrieved from http://adsabs.harvard.edu/abs/2009AGUFM.C42A.08V

70. **Larsen, C. F. et al. (2004).** Rapid uplift of southern Alaska caused by recent ice loss. *Geophysical Journal International,* 158(3), 1118-1133.

71. **Traufetter, G. (12-2-2010).** Sea Level Could Rise in South, Fall in North. *Spiegel Online,*. Retrieved from http://www.spiegel.de/international/world/0,1518,732303,00.html

72. **Buis, A. (8-23-2011).** NASA Satellites Detect Pothole on Road to Higher Seas; Global Sea Level Drops 6mm in 2010. Retrieved from http://www.jpl.nasa.gov/news/news.php?release=2011-262

73. **Doran, P. T., & Zimmerman, M. K. (2009).** Direct examination of the scientific consensus on climate change. *EOS,* 90 (3), 22.

74. **Scientific Consensus and the myth of 31,000 scientists disputing human connection.** Retrieved from http://www.skepticalscience.com/global-warming-scientific-consensus.htm

75. **Muller, R. A. (7-28-2012)**. The Conversion of a Climate-Change Skeptic. *New York Times*, Retrieved from http://www.nytimes.com/2012/07/30/opinion/the-conversion-of-a-climate-change-skeptic.html?pagewanted=all

76. **Hansen, J. E. (2007)**. Scientific reticence and sea level rise. *Environmental Research Letters*, 2, 024002. Retrieved from http://iopscience.iop.org/1748-9326/2/2/024002

77. **Ward, P. D. (2008)**. *Under a green sky: global warming, the mass extinctions of the past, and what they can tell us about our future*. Harper Paperbacks.

78. **Weitzman, M. (2010)**. GHG targets as insurance against catastrophic climate damages. *NBER Working Paper*. Retrieved from http://papers.ssrn.com/sol3/papers.cfm?abstract_id=1630141

79. **Zachos, J. C. et al. (2008)**. An early Cenozoic perspective on greenhouse warming and carbon-cycle dynamics. *Nature*, 451(7176), 279-283.

80. **Hansen, J. E. (2009)**. *Storms of my grandchildren : the truth about the coming climate catastrophe and our last chance to save humanity*. (1st U.S. ed. ed.). New York: Bloomsbury USA.

81. **Boucher, O. et al. (2009)**. The indirect global warming potential and global temperature change potential due to methane oxidation. *Environmental Research Letters*, 4(4), 044007.

82. **Shindell, D. T. et al. (2009)**. Improved attribution of climate forcing to emissions. *Science*, 326(5953), 716.

83. **Thomas, D. J. et al. (2002)**. Warming the fuel for the fire: Evidence for the thermal dissociation of methane hydrate during the Paleocene-Eocene thermal maximum. *Geology*, 30(12), 1067-1070. doi:10.1130

84. **Katz, M. E. et al. (2001)**. Uncorking the bottle: What triggered the Paleocene/Eocene thermal maximum methane release. *Paleoceanography*, 16(6), 549-562. Retrieved from http://geology.rutgers.edu

85. **Schaefer, K. et al. (2011)**. Amount and timing of permafrost carbon release in response to climate warming. *Tellus B*,

86. **Shakhova, N. et al. (2010)**. Extensive methane venting to the atmosphere from sediments of the East Siberian Arctic shelf. *Science*, 327(5970), 1246.

87. **Davidson, M. (6-10-2010)**. A Transect of topography in Charleston, SC from the tip of White Point Gardens up to Wentworth Street, between Smith and Pitt.

88. **Dewberry, I. (3-29-2012)**. National Enhanced Elevation Assessment. USGS. Retrieved from http://www.dewberry.com/files/pdf/NEEA_Final%20Report_Revised%203.29.12.pdf

89. **Strauss, B. (8-30-2011)**. New York's One-Inch Escape From Hurricane Irene. Retrieved from http://www.climatecentral.org/blogs/new-yorks-one-inch-escape-from-hurricane-irene/

90. **Group, SFRCCCSC. (2011)**. A Unified Sea Level Rise Projection for Southeast Florida. Fort Lauderdale: Southeast Florida Regional Climate Change Compact Steering Committee. www.broward.org/NaturalResources/ClimateChange/

91. **Bender, M. A. et al. (2010)**. Modeled impact of anthropogenic warming on the frequency of intense Atlantic hurricanes. *Science*, 327(5964), 454.

92. **Blake, E. et al. (2011)**. The Deadliest, Costliest, and Most Intense United StatesTropical Cyclones from 1851 - 21010. Miami: NOAA, National Weather Service, National Hurricane Center. http://www.nhc.noaa.gov/pdf/nws-nhc-6.pdf

93. **Webster, P. J. et al. (2005)**. Changes in tropical cyclone number, duration, and intensity in a warming environment. *Science*, 309(5742), 1844.

94. **Burkett, V. R. et al. (2002)**. Sea-level rise and subsidence: implications for flooding in New Orleans, Louisiana. Proceedings from US Geological Survey Subsidence Interest Group Conference: Proceedings of the Technical Meeting, Galveston, Texas, 27-29.

95. **Nicholls, R. J. et al. (2007)**. Ranking Port Cities with High Exposure and Vulnerability to Climate Extremes. *Organisation for Economic Co-operation and Development, Environment Working Paper*, 1, 53-57. Retrieved from http://www.aia.org/aiaucmp/groups/aia/documents/pdf/aias076737.pdf

96. **Obeysekera, J. (2009)**. Climate Change. Retrieved from https://my.sfwmd.gov

97. **Reid, A. (9-12-2011)**. Saltwater seeps into well waters in Hallandale Beach. *Sun Sentinel*, Fort Lauderdale

98. **Heimlich, B. N. et al. (2009).** South Florida's Resilient Water Resources: Adaptation to Sea Level Rise and Other Climate Changes. Retrieved from http://www.ces.fau.edu/files/projects/climate_change/SE_Florida_Resilient_Water_Resources.pdf

99. **Guinotte, J. M., & Fabry, V. J. (2008).** Ocean acidification and its potential effects on marine ecosystems. *Annals of the New York Academy of Sciences*, 1134(1), 320-342.

100. **Pandolfi, J. M. et al. (2011).** Projecting Coral Reef Futures Under Global Warming and Ocean Acidification. *Science*, 333(6041), 418-422.

101. **Barton, A. et al. (2012).** The Pacific oyster, Crassostrea gigas, shows negative correlation to naturally elevated carbon dioxide levels: Implications for near-term ocean acidification effects. *Limnol. Oceanogr*, 57(3), 698-710.

102. **Kiessling, W., & Simpson, C. (2011).** On the potential for ocean acidification to be a general cause of ancient reef crises. *Global Change Biology*, 17(1), 56–67.

103. **Sale, P. (2011).** *Our Dying Planet: An Ecologist's View of the Crisis We Face.* University of California Press.

104. **Department of Defense. (2010).** Quadrennial defense review report. Retrieved from http://www.defense.gov/qdr/images/QDR_as_of_12Feb10_1000.pdf

105. **Goodman, S. (2007).** National Security and the Threat of Climate Change. Retrieved from http://handle.dtic.mil/100.2/ADA469156

106. **Ward, P. D. (2010).** *The Flooded Earth: Our Future in a World Without Ice Caps.* New York: Basic Books.

107. **Lenton, T. et al. (2009).** Major Tipping Points in the Earth, Climate System and Consequences for the Insurance Sector. Tyndall Centre for Climate Change Research.

108. **Williamson, D. (3-9-2011).** Manager, Geographic Informaton Systems, New York City Dept of Information Technology. Personal communication.

109. **USGS.** Highest Natural Point of Elevation on the island of Manhattan. Retrieved from http://wikimapia.org/1667671/Highest-Natural-Point-of-Elevation-on-the-island-of-Manhattan-265-5-feet-USGS.

110. Larson, E. (9-25-1999). Hurricanes on the Hudson. *New York Times.* Retrieved from http://www.nytimes.com/1999/09/25/opinion/hurricanes-on-the-hudson.html

111. Kaufman, L. (11-26-10). Front-Line City in Virginia Tackles Rise in Sea. *New York Times,* p. 1. Retrieved from http://www.nytimes.com/2010/11/26/science/earth/26norfolk.html

112. Tullis, P. California's Delta Water Blues. Pacific Standard. Retrieved from http://www.psmag.com/environment/californias-delta-water-blues-26552/

113. Balica, S. et al. (2012). A flood vulnerability index for coastal cities and its use in assessing climate change impacts. *Natural Hazards,* 64, 73-105.

114. Lim, K. (2012). Shanghai denies it is most prone to flooding. *Channel News Asia.* Retrieved from http://www.channelnewsasia.com/stories/eastasia/view/1222627/1/.html

115. *New York Times* (2008). "Dutch draw up drastic measures to defend coast against rising seas." *New York Times.* Retrieved from http://www.nytimes.com/2008/09/03/news/03iht-03dutch.15877468.html

116. Dasgupta, S. (2007). *The impact of sea level rise on developing countries: A comparative analysis* (4136). World Bank Publications.

117. UN News Centre (2011). Palau Seeks UN World Court opinion on damage caused by greenhouse gases. Retrieved from http://www.un.org/en/

118. Rubin, J. (2010). International Law and the Victims of Climate Change: Creating a Framework for Managing Impacts and Displaced People. American Security Project. Retrieved from http://americansecurityproject.org/

119. Meakins, B. (9-5-2012). I-Kiribati Man Wanting 'Climate Change Refugee' Status Denied by New Zealand Immigration. *Huffington Post.* Retrieved from http://www.huffingtonpost.com/brook-meakins/kiribati-wanting-climate-_b_1859518.html

120. Domains. (5-4-2007).TV is Turned On...Again. Retrieved from http://www.dailydomainer.com/2007154-tv-is-turned-on-again.html

121. IMF. (2011). Tuvalu: 2010 Article IV Consultation—Staff Report. IMF Country Report No. 11/46. Retrieved from https://www.imf.org/external/pubs/ft/scr/2011/cr1146.pdf

122. **NFIP.** National Flood Insurance Program Statistics as of June 2012. Retrieved from http://common-resources.org/tag/nfip/

123. **Heberger, M. et al. (2009).** The Impacts of Sea-level Rise on the California Coast. California Climate Change Center.

124. **Van Raalten, D. (2009).** San Francisco Bay: Preparing for the Next Level. San Francisco: San Francisco Bay Conservation and Development Commission. Retrieved from http://www.bcdc.ca.gov/planning/climate_change/SFBay_preparing_%20for_the_next_Level.pdf

125. **Lynch, P. (2007).** Are Our Bases AT Risk from Rising Seas? *Daily Press*. Retrieved from www.dailypress.com

126. **Smetacek, V. et al. (2012).** Deep carbon export from a Southern Ocean iron-fertilized diatom bloom. *Nature*, 487(7407), 313-319.

127. **Ajai, L. B., et al. (2011).** Fate of Mountain Glaciers in the Anthropocene. Proceedings from Working Group of the Pontifical Academy of Sciences, Vatican City.

128. **Leiserowitz, A. (2012).** Global Warming's Six Americas in March 2012 and November 2011. Retrieved from http://environment.yale.edu/climate/files/Six-Americas-March-2012.pdf

129. **Churchill, S. W. (1940).** Speech to the British House of Commons regarding War Situation. *Hansard*, 362, section 60. Retrieved from http://hansard.millbanksystems.com/commons/1940/jun/18/war-situation#column_60

130. **Miller, K. G. et al. (2005/11/25).** The Phanerozoic Record of Global Sea-Level Change. Science, 310(5752), 1293-1298. doi:10.1126/science.1116412

Figures, Illustrations and Photos

Figure 6-3: Ranges of Project Sea Level Rise. Data from Rahmstorf, 2010

Page 45: Sharp's Lighthouse. Photo by Constantine M. Frangos

Page 46: Last House on Holland Island. Photo from flickr.com, baldeaglebluff

Figure 7-1: Greenland Ice Extent, 2004. Graph, courtesy of ACIA/ Map, Clifford Grabhorn

Figure 7-2: Moulin Diagram. Graph, courtesy of NASA

Page 55: Ililussat (Jakobshavn) Glacier Greenland. Photo from Google Earth

Page 55: Aerial view Greenland. Photo by John Englander

Figure 7-3: Potential Sea Level Rise Table. Adapted from Allison, *et al.*, 2009

Figure 7-4: East Antarctic Map, courtesy of Springer Science+Business Media.

Figure 7-5: Greenland Ice Melt. Photo, courtesy of the National Snow and Ice Data Center, University of Colorado, Boulder

Figure 7-6: Average Extent Graph. Image, courtesy of the National Snow and Ice Data Center, University of Colorado, Boulder

Figure 8-1: Sea Level, Temperature, and Carbon Dioxide over 400,000 Years. Graphic, courtesy of Hansen and Sato

Figure 8-2: Atmospheric $C0_2$ at Mauna Loa Observatory.Graphic, courtesy of esrl.noaa.gov/gmd/obop/mlo/

Figure 8-3: $C0_2$ Over Past 420,000 Years. Graphic, courtesy of NOAA

Page 70: Attributions of Climate Change. Graphic, courtesy of Lean and Rind

Figure 9-1: Global Warming Scenarios from IPCC 2013, Fifth Assessment Report. Graph, courtesy of Cambridge University Press.

Figure 9-2: Table of Sea Level Rise Projections this century adapted by author from Table SPM.2 in IPCC 2013 Assessment Report published by Cambridge University Press.

Figure 9-3: Mean Sea Level Graph. Graph, courtesy of S. Nerem, University of Colorado

Figure 10-1: Average Monthly Extent with additional trend line of recent years. Original graph, courtesy of the National Snow and Ice Data Center, University of Colorado, Boulder

Figure 10-2: Observed Versus Model Sea Ice Extent. Graph, courtesy of the National Snow and Ice Data Center, University of Colorado, Boulder

Figure 10-3: Comparison of two decades of SLR to IPCC projections. From Rahmstorf, "Comparing climate projections to observations up to 2011" originally published in Environmental Research Letters, 2012

Page 96: Methane from Ice. Photo, courtesy of Katey Walter Anthony's research team, University of Alaska Fairbanks.

Figure 10-4: Topographic Versus Lidar Graph. Graph, courtesy of Coastal Services Center, NOAA

Page 103: House for Sale, by John Englander

Figure 11-1: Broward County Map. Image adapted from ESRI / Dr. Lin Wu

Figure 11-2: Percentage of Hurricanes. Graph, courtesy of the Union of Concerned Scientists

Figure 11-3: Port City Sea Level Exposure Table.

Figure 12-1: Providence Map, from Wiki Commons

Figure 12-2: Washington, D.C. Map. Graph, courtesy of D.C. Department of Environment

Page 128: Venice Photo, from Wiki Commons

Figure 12-3: Populations at Risk. Graph, courtesy of Washington Center for Global Development

Page 159: Delta Works. Photo, courtesy of DeltaWorks Online, Job van de Sande

Figure 15-1: San Francisco Bay Plan. Graph, courtesy of San Francisco Bay Conservation and Development Commission

Figure 15-2: Boston Harbor. Graph, courtesy of Di Mambro + Associates

Index:

Acknowledgments

Many helped make this book a reality. It crystalized during a 2007 trip to Greenland sponsored by Charlie Gallagher. Leading Arctic expert Robert "Bob" Corell was our scientist-guide and became an invaluable advisor for this book. Bob is well known globally for his sage guidance and seemingly tireless efforts to advance the science, the policy, and the communication about climate change.

Dozens of leading scientists have been generous with their time to explain nuances. I can not possibly name them all, but each has my sincere appreciation for their office time, e-mails, and encouragement. More important I want to acknowledge the crucial work they are doing to advance the science that underlies the story of sea level rise and climate change.

To my mind those two related issues are nothing less than the greatest single challenge for mankind this century. Scientists, advocates, and some in government are advancing understanding and action on this essential issue. We should all recognize their contribution to future generations.

In particular I have to thank Dr. James Hansen. Amazingly, he finds time to advance the frontiers of climate physics, to be a passionate globe-trotting spokesman on the issue, and also to respond to my rather trivial questions. He is the definition of a hero for this planet.

Several senior people at the National Oceanic and Atmospheric Administration provided particular help and encouragement: Sandy MacDonald and his team at the Earth Systems Research Lab; Margaret Davidson and staff at the Coastal Services Center; and Rick Spinrad, formerly head of Ocean and Atmospheric Research.

John Jo Lewis and Jay Wade of subaviators.com provided valuable assistance with the Hawaii exploration of ancient coastlines. Len Berry, Patti Berry, Sylvia Earle, Gary Griggs, Richard Hildreth, Judy Keiser, Ken Nemeth and Hal Wanless carefully read the manuscript and offered excellent suggestions for its improvement. Matt Celestino, a geologist, developed many of the profiles of city vulnerabilities. My colleagues at the Global Environment and Technology Foundation have been an incredible resource.

A number of long-time friends had very relevant experience and were generous with valuable advice along the way: Peter Emerson, Mike Emmerman, Capt. Joel Fogel, Bill Gleason, Mark Grosvenor, David James, Daniel Kreeger, Paula Lowe, Ken Loyst, John Nightingale, Neil Ross, Remar Sutton, and Peter Vassilopoulos.

I was incredibly lucky to find editor Jill Buchanan, who brought skill, insight, and passion making this book much better than it would have been otherwise. Kata Jancso, my cover artist from Budapest, is another amazing talent. Roberta Stealy was always there to help with the dozens of illustrations. My assistant Sharon Gray has been a great asset. When Sandy hit a week after the book was published, Paul Krupin handled media relations and proved to have all the right experience.

The first edition generated hundreds of contacts from readers who have provided wonderful comments and encouragement. A few have become friends and colleagues, generously helping me carry my message farther: Jamie Carson, Mitch Chester, Dave Finnigan, Ken Hirsch, Bill Morrison, Wayne Pathman, Rebecca Rubin and Will Travis.

Linda, my wonderful wife of 20 years, surprised me with her extraordinary editorial expertise. She and Rachel, our daughter, gave me great support during the seemingly endless months working, or obsessing, on this projet.

Thank you one and all, and to those I have not individually identified. I am truly grateful. Not only does this book allow me to share what I have learned, but writing it greatly advanced my understanding as well.

About the author

John Englander is an oceanographer, consultant and author. He has witnessed first hand the impacts of climate change, through expeditions under the polar ice cap, deep dives in research submarines and visits to Greenland and Antarctica. A broad marine science background coupled with degrees in Geology and Economics allows him to see the big picture on climate and look ahead to the large-scale financial and societal impacts, particularly as they relate to sea level rise.

For over 30 years, Englander has been a leader in both the private sector and the non-profit arena, serving as CEO for such noteworthy organizations as The Cousteau Society and The International SeaKeepers Society.

Englander works with businesses, government agencies, and communities to understand the financial risks as increasing severe storms and long term sea level rise challenge us to adapt to a shoreline that will move inland for decades.

Recently Mr. Englander has been in the media worldwide, appearing on MSNBC, National Public Radio (NPR), the FOX Business Channel and SkyNews TV-UK. He has been featured in USA Today, Huffington Post, the San Francisco Chronicle, and Publisher's Weekly.

John Englander is a Fellow of the Institute of Marine Engineering, Science and Technology, and The Explorers Club. He is the Special Advisor on Climate to Friends of the United Nations and a member of the American Geophysical Union, the American Association for the Advancement of Science, the Union of Concerned Scientists, and the Marine Technology Society.

Made in the USA
Charleston, SC
24 February 2017